CAMBRIDGE TRACTS IN MATHEMATICS

General Editors

B. BOLLOBÁS, W. FULTON, A. KATOK, F. KIRWAN,
P. SARNAK, B. SIMON, B. TOTARO

184 Algebraic Theories

Algebraic Theories
A Categorical Introduction to General Algebra

J. ADÁMEK
Technische Universität Carolo Wilhelmina zu Braunschweig, Germany

J. ROSICKÝ
Masarykova Univerzita v Brně, Czech Republic

E. M. VITALE
Université Catholique de Louvain, Belgium

With a Foreword by F. W. LAWVERE

CAMBRIDGE UNIVERSITY PRESS
Cambridge, New York, Melbourne, Madrid, Cape Town, Singapore,
São Paulo, Delhi, Dubai, Tokyo, Mexico City

Cambridge University Press
The Edinburgh Building, Cambridge CB2 8RU, UK

Published in the United States of America by Cambridge University Press, New York

www.cambridge.org
Information on this title: www.cambridge.org/9780521119221

© J. Adámek, J. Rosický, and E. M. Vitale 2011
Foreword © F. W. Lawvere 2011

This publication is in copyright. Subject to statutory exception
and to the provisions of relevant collective licensing agreements,
no reproduction of any part may take place without the written
permission of Cambridge University Press.

First published 2011

Printed in the United Kingdom at the University Press, Cambridge

A catalogue record for this publication is available from the British Library

Library of Congress Cataloguing in Publication data
Adámek, Jiří, dr.
Algebraic theories : a categorical introduction to general algebra / J. Adámek,
J. Rosický, E. M. Vitale.
p.   cm. – (Cambridge tracts in mathematics ; 184)
Includes bibliographical references and index.
ISBN 978-0-521-11922-1 (hardback)
1. Categories (Mathematics)   2. Algebraic logic.   I. Rosický, Jiří.   II. Vitale, E. M.
III. Title.   IV. Series.
QA169.A31993   2010
512'.62 – dc22   2010018289

ISBN 978-0-521-11922-1 Hardback

Cambridge University Press has no responsibility for the persistence or
accuracy of URLs for external or third-party internet websites referred to
in this publication, and does not guarantee that any content on such
websites is, or will remain, accurate or appropriate.

To Susy, Radka, and Ale

Contents

Foreword		*page* ix
F. W. Lawvere		
Preface		xv

PART I: ABSTRACT ALGEBRAIC CATEGORIES

0	Preliminaries	3
1	Algebraic theories and algebraic categories	10
2	Sifted and filtered colimits	21
3	Reflexive coequalizers	30
4	Algebraic categories as free completions	38
5	Properties of algebras	46
6	A characterization of algebraic categories	54
7	From filtered to sifted	65
8	Canonical theories	74
9	Algebraic functors	80
10	Birkhoff's variety theorem	89

PART II: CONCRETE ALGEBRAIC CATEGORIES

11	One-sorted algebraic categories	103
12	Algebras for an endofunctor	117

13	Equational categories of Σ-algebras	127
14	*S*-sorted algebraic categories	139

PART III: SPECIAL TOPICS

15	Morita equivalence	153
16	Free exact categories	163
17	Exact completion and reflexive-coequalizer completion	182
18	Finitary localizations of algebraic categories	195
	Postscript	204
Appendix A	Monads	207
Appendix B	Abelian categories	227
Appendix C	More about dualities for one-sorted algebraic categories	232
	References	241
	List of symbols	245
	Index	247

Foreword

The study Birkhoff initiated in 1935 was named *general algebra* in Kurosh's classic text; the subject is also called *universal algebra*, as in Cohn's text. The purpose of general algebra is to make explicit common features of the practices of commutative algebra, group theory, linear algebra, Lie algebra, lattice theory, and so on, to illuminate the paths of those practices. In 1963, less than 20 years after the 1945 debut of the Eilenberg–Mac Lane method of categorical transformations, its potential application to general algebra began to be developed into concrete mathematical practice, and that development continues in this book.

Excessive iteration of the passage

$$\mathcal{T}' = \text{theory of } \mathcal{T}$$

would be sterile if pursued as idle speculation without attention to that fundamental motion of theory: concentrate the essence of practice to guide practice. Such theory is necessary to clear the way for the advance of teaching and research. General algebra can and should be used in particular algebras (i.e., in algebraic geometry, functional analysis, homological algebra, etc.) much more than it has been. There are several important instruments for such application, including the partial structure theorem in Birkhoff's *Nullstellensatz*, the *commutator* construction, and the general framework itself.

Birkhoff's theorem was inspired by theorems of Hilbert and Noether in algebraic geometry (as indeed was the more general model theory of Robinson and Tarski). His greatest improvement was not only in generality: beyond the mere existence of generalized points, he showed that they are sufficient to give a monomorphic embedding. Nevertheless, in commutative algebra his result is rarely mentioned (although it is closely related to Gorenstein algebras). The categorical formulation of Birkhoff's theorem (Lawvere, 2008; Tholen, 2003), like precategorical formulations, involves subdirect irreducibility and

Zorn's lemma. Finitely generated algebras in particular are partially dissected by the theorem into (often qualitatively simpler) finitely generated pieces. For example, when verifying consequences of a system of polynomial equations over a field, it suffices to consider all possible finite-dimensional interpretations, where constructions of linear algebra such as the trace construction are available.

Another accomplishment of general algebra is the so-called commutator theory (named for its realization in the particular category of groups); a categorical treatment of this theory can be found in the work of Pedicchio (1995) and Janelidze and Pedicchio (2001). In other categories this theory specializes to a construction important in algebraic geometry and number theory, namely, the product of ideals (Hagemann & Hermann, 1979). In the geometrical classifying topos for the algebraic category of K-rigs, this construction yields an internal multiplicative semilattice of closed subvarieties.

In the practice of group theory and ring theory, the roles of presentations and of the algebras presented have long been distinguished, giving a syntactic approach to calculation, in particular algebraic theories. Yet many works in general algebra (and model theory generally) continue anachronistically to confuse a presentation in terms of signatures with the presented theory itself, thus ignoring the application of general algebra to specific theories, such as that of C^∞-rings, for which no presentation is feasible.

Apart from the specific accomplishments mentioned previously, the most effective illumination of algebraic practice by general algebra, both classical and categorical, has come from the explicit nature of the framework itself. The closure properties of certain algebraic subcategories, the functorality of semantics itself, the ubiquitous existence of functors adjoint to algebraic functors, and the canonical method for extracting algebraic information from nonalgebraic categories have served (together with their many particular ramifications) as partial guidance to mathematicians dealing with the inevitably algebraic content of their subjects. The careful treatment of these basics by Adámek, Rosický, and Vitale will facilitate future mutual applications of algebra, general algebra, and category theory. The authors have achieved in this book the new resolution of several issues that should lead to further research.

What is general algebra?

The bedrock ingredient for all of general algebra's aspects is the use of finite Cartesian products. Therefore, as a framework for the subject, it is appropriate to recognize the 2-category of categories that have finite categorical products and of the functors preserving these products. Among such categories are linear categories whose products are simultaneously coproducts; this is a crucial property of linear algebra in that maps between products are then uniquely represented as matrices of smaller maps between the factors (though of course, there is no unique decomposition of objects into products, so it would be incorrect to say inversely that maps "are" matrices). General categories with products can be forced to become linear, and this reflection 2-functor is an initial ingredient in linear representation theory. However, I want to emphasize instead a strong analogy between general algebra as a whole and any particular linear monoidal category because that will reveal some of the features of the finite product framework that make the more profound results possible.

The 2-category of all categories with finite products has (up to equivalence) three characteristic features of a linear category such as the category of modules over a rig:

1. It is *additive* because if $A \times B$ is the product of two categories with finite products, it is also their coproduct, the evident injections from A, B having the universal property for maps onto any third such category.
2. It is *symmetric closed*; indeed Hom(A, B) is the category of algebras in the background B according to the theory A. The unit I for this Hom is the opposite of the category of finite sets. The category J of finite sets itself satisfies Hom(J, J) = I, and the category Hom(J, B) is the category of Boolean algebras in B. As dualizer, the case in which B is the category of small sets is most often considered in abstract algebra.
3. It is *tensored* because a functor of two variables that is product preserving in each variable separately can be represented as a product-preserving functor on a suitable tensor product category. Such functors occur in the recent work of Janelidze (2006); specifically, there is a canonical evaluation $A \otimes$ Hom(A, B) $\to B$, where the domain is "a category whose maps involve both algebraic operations and their homomorphisms."

A feature not present in abstract linear algebra (though it has an analog in the cohesive linear algebra of functional analysis) is Street's *bo-ff factorization* of any map (an abbreviation for "bijective on objects followed by full and faithful"; see Street, 1974; Street & Walters, 1978). A single-sorted algebraic theory is a map $I \to A$ that is bijective on objects; such a map induces a single "underlying" functor Hom(A, B) $\to B$ on the category of A algebras in B. The factorization permits the definition of the full "algebraic structure" of any given map $u: X \to B$, as follows: the map $I \to$ Hom(X, B) that represents u has its bo-ff factorization, with its bo part being the algebraic theory $I \to A(u)$, the full X-natural structure (in its abstract general guise) of all values of u. The original u lifts across the canonical $\mathrm{Hom}(A(u), B) \to B$ by a unique $u^{\#}$. This is a natural first step in one program for "inverting" u because if we ask whether an object of B is a value of u, we should perhaps consider the richer (than B) structure that any such object would naturally have; that is, we change the problem to one of inverting $u^{\#}$. Jon Beck called this program "descent" with respect to the "doctrine" of general algebra. (A second step is to consider whether $u^{\#}$ has an adjoint.)

Frequently, the dualizing background B is Cartesian closed; that is, it has not only products but also finite coproducts and exponentiation, where exponentiation is a map

$$B^{op} \otimes B \to B$$

in our 2-category. This permits the construction of the important family of function algebras $B^{op} \to \mathrm{Hom}(A, B)$ given any A algebra (of "constants") in B.

On a higher level, the question whether a given C is a value of the 2-functor $U = \mathrm{Hom}(-, B)$ (for given B) leads to the discovery that such values belong to a much richer doctrine, involving as operations all limits that B has and all colimits that exist in B and preserve finite products. As in linear algebra, where dualization in a module B typically leads to modules with a richer system of operators, conversely, such a richer structure assumed on C is a first step toward 2-descent back along U.

The power of the doctrine of natural 2-operations on $\mathrm{Hom}(-, B)$ is enhanced by fixing B to be the category of small sets, where smallness specifically excludes measurable cardinals (although they may be present in the categorical universe at large).

A contribution of Birkhoff's original work had been the characterization of varieties, that is, of those full subcategories of a given algebraic category $\mathrm{Hom}(A, B)$ that are equationally defined by a surjective map $A \to A'$ of theories. Later, the algebraic categories themselves were characterized. Striking refinements of those characterization results, in particular the clarification of a question left open in the 1968 treatment of categorical general algebra (Lawvere, 1969), are among the new accomplishments explained in this book. As Grothendieck had shown in his very successful theory of Abelian categories, the exactness properties found in abstract linear algebra continue to be useful for the concretely variable linear algebras arising as sheaves in complex analysis; should something similar be true for nonlinear general algebras? More specifically, what are the natural 2-operations and exactness properties shared by all the set-valued categories concretely arising in general algebra? In particular, can that class of categories be characterized by further properties, such as sufficiency of projectives, in terms of these operations? It was clear that small limits and filtered colimits were part of the answer, as with the locally finitely presentable categories of Gabriel and Ulmer. But the further insistence of general algebra on algebraic operations that are total leads to a further functorial operation, needed to isolate equationally the correct projectives and that is also useful in dealing with non-Mal'cev categories – that further principle is the ubiquitous preservation of Linton's reflexive coequalizers, which are explained in this book as a crucial case of Lair's sifted colimits.

F. W. Lawvere

Preface

F. W. Lawvere's introduction of the concept of an algebraic theory in 1963 proved to be a fundamental step toward developing a categorical view of general algebra in which varieties of algebras are formalized without details of equational presentations. An algebraic theory as originally introduced is, roughly speaking, a category whose objects are all finite powers of a given object. An algebra is then a set-valued functor preserving finite products, and a homomorphism between algebras is a natural transformation. In the almost half a century that has followed Lawvere's introduction, this idea has gone through a number of generalizations, ramifications, and applications in areas such as algebraic geometry, topology, and computer science. The generalization from one-sorted algebras to many-sorted algebras (of particular interest in computer science) leads to a simplification: an algebraic theory is now simply a small category with finite products.

Abstract algebraic categories

In Part I of this book, consisting of Chapters 1–10, we develop the approach in which algebraic theories are studied without reference to sorting. Consequently, algebraic categories are investigated as abstract categories. We study limits and colimits of algebras, paying special attention to the sifted colimits because they play a central role in the development of algebraic categories. For example, algebraic categories are characterized precisely as the free completions under sifted colimits of small categories with finite coproducts, and algebraic functors are precisely the functors preserving limits and sifted colimits. This leads to an algebraic duality: the 2-category of algebraic categories is dually biequivalent to the 2-category of canonical algebraic theories.

Here we present the concept of equation as a parallel pair of morphisms in algebraic theory. An algebra satisfies the equation iff it merges the parallel pair. We prove Birkhoff's variety theorem: subcategories that can be presented by equations are precisely those that are closed under products, subalgebras, regular quotients, and directed unions. (The last item can be omitted in case of one-sorted algebras.)

Concrete algebraic categories

Lawvere's original concept of one-sorted theory is studied in Chapters 11–13. Here the categories of algebras are concrete categories over *Set*, and we prove that up to concrete equivalence, they are precisely the classical equational categories of Σ-algebras for one-sorted signatures Σ. More generally, given a set S of sorts, we introduce in Chapter 14 S-sorted algebraic theories and the corresponding S-sorted algebraic categories that are concrete over S-sorted sets. Thus we distinguish between *many-sorted* algebras, where sorting is not specified, and *S-sorted* algebras, where a set S of sorts is given (and this distinction leads us to consider categories of algebras as concrete or abstract).

This discussion is supplemented by Appendix A, in which we present a short introduction to monads and monadic algebras. In Appendix C we prove a duality between one-sorted algebraic theories and finitary monadic categories over *Set* and again, more generally, between S-sorted algebraic theories and finitary monadic categories over Set^S.

Abelian categories are shortly treated in Appendix B.

The non-strict versions of some concepts, such as morphism of one-sorted theories and concrete functors, are treated in Appendix C.

Special topics

Chapters 15–18 are devoted to some more specialized topics. Here we introduce Morita equivalence, characterizing pairs of algebraic theories yielding equivalent categories of algebras. We also prove that algebraic categories are free exact categories. Finally, the finitary localizations of algebraic categories are described: they are precisely the exact locally finitely presentable categories.

Other topics

Of the two categorical approaches to general algebra, monads and algebraic theories, only the latter is treated in this book, with the exception of the short appendix on monads. Both of these approaches make it possible to study algebras in a general category; in our book we just restrict ourselves to sets and many-sorted sets. Thus, examples such as topological groups are not treated here.

Other topics related to our book are mentioned in the Postscript.

Interdependance of chapters

Until Chapter 7 inclusive every chapter is strongly dependent on the previous ones. But some topics in the sequel of the book can be studied by skipping Chapters 2–7 (and consulting them just for specific definitions and results):

    algebraic duality in Chapters 8–9
    Birkhoff's variety theorem in Chapter 10
    one-sorted theories in Chapters 11–13
    S-sorted theories in Chapter 14 (after reading Chapter 11)
    Morita equivalence of theories in Chapter 15

Acknowledgments

The authors are indebted to Bill Lawvere who had the idea that a book about algebraic theories was needed, and who followed the years of the development of our project with great interest and with critical comments that helped us immensely. Also discussions with Walter Tholen in the early stages of our book were of great value. We are also grateful to a number of colleagues for their suggestions on the final text, particularly to Michel Hébert, Alexander Kurz, Raul Leal, Stefan Milius, John Power and Ross Street.

PART I

Abstract algebraic categories

It should be observed first that the whole concept of category is essentially an auxiliary one; our basic concepts are those of a functor and a natural transformation.
– S. Eilenberg and S. Mac Lane (1945), General theory of natural equivalences, *Trans. Amer. Math. Soc.* **58**: 247

0
Preliminaries

The aim of this chapter is to fix some notation and recall well-known facts concerning basic concepts of category theory used throughout the book. The reader may well skip it and return to it when needed. Only the most usual definitions and results of the theory of categories are mentioned here; more about them can be found in any of the books mentioned at the end of this chapter.

0.1 Foundations In category theory, one needs to distinguish between *small* collections (sets) and *large* ones (classes). An arbitrary set theory making such a distinction possible is sufficient for our book. The category of (small) sets and functions is denoted by

$$Set.$$

All categories with which we work have small hom-sets. It follows that every object has only a set of retracts (see 0.16) up to isomorphism.

0.2 Properties of functors A functor $F\colon \mathcal{A} \to \mathcal{B}$ is

1. *faithful* if for every parallel pair of morphisms $f, g\colon A \rightrightarrows A'$ in $\mathcal{A}$, one has $f = g$ whenever $Ff = Fg$
2. *full* if for every morphism $b\colon FA \to FA'$ in $\mathcal{B}$, there exists a morphism $a\colon A \to A'$ in $\mathcal{A}$ such that $Fa = b$
3. *essentially surjective* if for every object B in $\mathcal{B}$, there exists an object A in $\mathcal{A}$ with B isomorphic to FA
4. an *equivalence* if there exists a functor $F'\colon \mathcal{B} \to \mathcal{A}$ such that both $F \cdot F'$ and $F' \cdot F$ are naturally isomorphic to the identity functors; such a functor F' is called a *quasi-inverse* of F

5. an *isomorphism* if there exists a functor $F'\colon \mathcal{B} \to \mathcal{A}$ such that both $F \cdot F'$ and $F' \cdot F$ are equal to the identity functors
6. *conservative* if it reflects isomorphisms; that is, $a\colon A \to A'$ is an isomorphism whenever $Fa\colon FA \to FA'$ is

0.3 Remark Let $F\colon \mathcal{A} \to \mathcal{B}$ be a functor.

1. If F is full and faithful, then it is conservative.
2. F is an equivalence iff it is full, faithful, and essentially surjective.
3. If F is an equivalence and F' a quasi-inverse of F, it is possible to choose natural isomorphisms $\eta\colon \mathrm{Id}_{\mathcal{B}} \to F \cdot F'$ and $\varepsilon\colon F' \cdot F \to \mathrm{Id}_{\mathcal{A}}$ such that

$$F\varepsilon \cdot \eta F = F \quad \text{and} \quad \varepsilon F' \cdot F'\eta = F'$$

see 0.8 below. (Observe that in equations like $F\varepsilon \cdot \eta F = F$, we write F for the identity natural transformation on a functor F. We adopt the same convention in diagrams having functors as vertices and natural transformations as edges.)
4. F is an isomorphism iff it is full, faithful, and bijective on objects.

0.4 Functor categories and Yoneda embedding

1. Given a category $\mathcal{A}$ and a small category $\mathcal{C}$, we denote by $\mathcal{A}^{\mathcal{C}}$ the category of functors from $\mathcal{C}$ to $\mathcal{A}$ and natural transformations.
2. In case $\mathcal{A} = Set$, we have the *Yoneda embedding*:

$$Y_{\mathcal{C}}\colon \mathcal{C}^{op} \to Set^{\mathcal{C}} \quad Y_{\mathcal{C}}(X) = \mathcal{C}(X, -),$$

which is full and faithful. This follows from the *Yoneda lemma*, which states that for every $X \in \mathcal{C}$ and for every functor $F\colon \mathcal{C} \to Set$, the map assigning to every natural transformation $\alpha\colon Y_{\mathcal{C}}(X) \to F$ the value $\alpha_X(\mathrm{id}_X)$ is a bijection natural in X and F.

0.5 Diagrams

1. A *diagram* in a category $\mathcal{K}$ is a functor from a small category into $\mathcal{K}$.
2. A *finite diagram* is a diagram $D\colon \mathcal{D} \to \mathcal{K}$ such that $\mathcal{D}$ is a *finitely generated category*. This means that $\mathcal{D}$ has finitely many objects and a finite set of morphisms whose closure under composition gives all the morphisms of $\mathcal{D}$.
3. A category is *complete* if limits of all diagrams in it exist and *cocomplete* if all colimits exist.

4. Limits commute with products: given a set C and a collection of diagrams

$$D_c: \mathcal{D} \to \mathcal{A}, \quad c \in C$$

in a complete category $\mathcal{A}$, the product of their limits is the same as the limit of their product. This can be formalized by viewing C as a discrete category and considering the limit of the obvious diagram $D: C \times \mathcal{D} \to \mathcal{A}$. The statement is that the canonical morphism from $\lim D$ to $\Pi(\lim D_c)$ is an isomorphism.

5. Limits commute with limits. This generalizes item 4 to the case when C is an arbitrary small category: given a complete category $\mathcal{A}$ and a diagram $D: C \times \mathcal{D} \to \mathcal{A}$, the canonical morphisms

$$\lim_{C} (\lim_{\mathcal{D}} D(c, d)) \leftrightarrows \lim_{\mathcal{D}} (\lim_{C} D(c, d))$$

are mutually inverse isomorphisms. Moreover, each one of these isomorphic objects is a limit of D.

0.6 Colimits in *Set* In the category of sets:

1. Coproducts are disjoint unions.
2. Coequalizers of $p, q: X \rightrightarrows Z$ can be described as the canonical maps $c: Z \to Z/\sim$, where $\sim$ is the smallest equivalence relation with $p(x) \sim q(x)$ for every $x \in X$. This equivalence relation merges elements z and z' of Z iff there exists a zigzag of elements

where each f_k is equal to p or q for $k = 0, \ldots, 2n + 1$.

3. A category $\mathcal{D}$ is called *filtered* if every finitely generated subcategory has a cocone in $\mathcal{D}$ (for more about this concept, see Chapter 2). Filtered diagrams are diagrams with a filtered domain. A colimit of a filtered diagram $D: \mathcal{D} \to Set$ is described as the quotient

$$\coprod_{x \in \mathrm{obj}\,\mathcal{D}} Dx / \sim$$

where for elements $u_i \in Dx_i$, we have $u_1 \sim u_2$ iff there exist morphisms $f_i: x_i \to y$ in $\mathcal{D}$ such that $Df_1(u_1) = Df_2(u_2)$.

0.7 Construction of colimits In a category $\mathcal{A}$ with coproducts and coequalizers, all colimits exist. Given a diagram $D\colon \mathcal{D} \to \mathcal{A}$ form a parallel pair

$$\coprod_{f \in \mathrm{mor}\,\mathcal{D}} D f_d \underset{j}{\overset{i}{\rightrightarrows}} \coprod_{x \in \mathrm{obj}\,\mathcal{D}} Dx$$

where f_d and f_c denote the domain and codomain of f. The f-component of i is the coproduct injection of $D f_d$; that of j is the composite of Df and the coproduct injection of $D f_c$.

1. If
$$c\colon \coprod_{x \in \mathrm{obj}\,\mathcal{D}} Dx \longrightarrow C$$

   is the coequalizer of i and j, then $C = \operatorname{colim} D$ and the components of c form the colimit cocone.

2. The preceding pair i, j is reflexive; that is, there exists a morphism
$$\delta\colon \coprod_{x \in \mathrm{obj}\,\mathcal{D}} Dx \longrightarrow \coprod_{f \in \mathrm{mor}\,\mathcal{D}} D f_d$$

   such that $i \cdot \delta = \mathrm{id} = j \cdot \delta$. Indeed, the x-component of δ is the coproduct injection of id_x.

0.8 Adjoint functors Given functors $U\colon \mathcal{A} \to \mathcal{B}$ and $F\colon \mathcal{B} \to \mathcal{A}$, then F is a left adjoint of U, with notation $F \dashv U$, if there exist natural transformations $\eta\colon \mathrm{Id}_\mathcal{B} \to UF$ and $\varepsilon\colon FU \to \mathrm{Id}_\mathcal{A}$ (called unit and counit) satisfying

$$\varepsilon F \cdot F\eta = F \quad \text{and} \quad U\varepsilon \cdot \eta U = U.$$

This is equivalent to the existence of a bijection

$$\mathcal{A}(FB, A) \simeq \mathcal{B}(B, UA)$$

natural in $A \in \mathcal{A}$ and $B \in \mathcal{B}$.

1. Every left adjoint preserves colimits.
2. Dually, every right adjoint preserves limits.
3. A *solution set* for a functor $U\colon \mathcal{A} \to \mathcal{B}$ and an object X of $\mathcal{B}$ is a set of morphisms $f_i\colon X \to UA_i$ $(i \in I)$ with $A_i \in \mathcal{A}$ such that every other morphism $f\colon X \to UA$ has a factorization $f = Uh \cdot f_i$ for some $i \in I$ and some morphism $h\colon A_i \to A$ in $\mathcal{A}$.
4. The *adjoint functor theorem* states that if $\mathcal{A}$ has limits, then a functor $U\colon \mathcal{A} \to \mathcal{B}$ has a left adjoint iff it preserves limits and has a solution set for every object X of $\mathcal{B}$.

0.9 Reflective subcategories Given a category $\mathcal{B}$, by a *reflective subcategory* of $\mathcal{B}$ is meant a subcategory $\mathcal{A}$ such that the inclusion functor $\mathcal{A} \to \mathcal{B}$ has a left adjoint (called a reflector for $\mathcal{B}$). We denote by $R: \mathcal{B} \to \mathcal{A}$ the reflector and by $r_B: B \to RB$ the reflections, that is, the components of the unit of the adjunction.

0.10 Representable functors A functor from a category $\mathcal{A}$ to *Set* is *representable* if it is naturally isomorphic to a hom-functor $\mathcal{A}(A, -)$.

1. If $\mathcal{A}$ has coproducts, then $\mathcal{A}(A, -)$ has a left adjoint assigning to a set X a coproduct of X copies of A.
2. The colimit of $\mathcal{A}(A, -)$ is a singleton set.
3. If $F: \mathcal{A} \to \mathit{Set}$ has a left adjoint then it is representable.

0.11 Examples of left adjoints

1. For every set X, the functor $X \times - : \mathit{Set} \to \mathit{Set}$ is left adjoint to $\mathit{Set}(X, -)$.
2. For a category $\mathcal{A}$, the diagonal functor

$$\Delta: \mathcal{A} \to \mathcal{A} \times \mathcal{A}, \quad A \mapsto (A, A)$$

has a left adjoint iff $\mathcal{A}$ has finite products. Then the functor

$$\mathcal{A} \times \mathcal{A} \to \mathcal{A}, \quad (A, B) \mapsto A \times B$$

is a left adjoint to Δ.

0.12 Remark The contravariant hom-functors $\mathcal{B}(-, B): \mathcal{B} \to \mathit{Set}^{op}$, $B \in \mathrm{obj}\,\mathcal{B}$, collectively reflect colimits; that is, for every cocone C of a diagram $D: \mathcal{D} \to \mathcal{B}$, we have the following: C is a colimit of D iff the image of C under any $\mathcal{B}(-, B)$ is a colimit of the diagram $\mathcal{B}(-, B) \cdot D$ in Set^{op}.

0.13 Slice categories Given functors $F: \mathcal{A} \to \mathcal{K}$ and $G: \mathcal{B} \to \mathcal{K}$, the *slice category* $(F \downarrow G)$ has as objects all triples (A, f, B) with $A \in \mathcal{A}$, $B \in \mathcal{B}$, and $f: FA \to GB$, and as morphisms $(A, f, B) \to (A', f', B')$, all pairs $a: A \to A'$, $b: B \to B'$ such that $Gb \cdot f = f' \cdot Fa$.

1. As special cases, we have $K \downarrow G$ and $F \downarrow K$, where an object $K \in \mathcal{K}$ is seen as a functor from the one-morphism category to $\mathcal{K}$.
2. If F is the identity functor on $\mathcal{K}$, we write $K \downarrow K$ instead of $\mathrm{id}_\mathcal{K} \downarrow K$.

0.14 Set functors as colimits of representables Every functor $A\colon \mathcal{T} \to Set$ ($\mathcal{T}$ small) is in a canonical way a colimit of representable functors. In fact, consider the Yoneda embedding $Y_\mathcal{T}\colon \mathcal{T}^{op} \to Set^\mathcal{T}$ and the slice category $El\, A = Y_\mathcal{T} \downarrow A$ of "elements of A." Its objects can be represented as pairs (X, x) with $X \in \mathrm{obj}\,\mathcal{T}$ and $x \in AX$, and its morphisms $f\colon (X, x) \to (Z, z)$ are morphisms $f\colon Z \to X$ of $\mathcal{T}$ such that $Af(z) = x$. We denote by $\Phi_A\colon El\,A \to \mathcal{T}^{op}$ the canonical projection that to every element of the set AX assigns the object X. Then A is a colimit of the following diagram of representable functors

$$El\,A \xrightarrow{\Phi_A} \mathcal{T}^{op} \xrightarrow{Y_\mathcal{T}} Set^\mathcal{T}.$$

Indeed the colimit injection $Y_\mathcal{T}(\Phi_A(X, x)) \to A$ is the natural transformation corresponding, by Yoneda lemma, to the element $x \in AX$.

0.15 Kernel pair A *kernel pair* of a morphism $f\colon A \to B$ is a parallel pair $f_1, f_2\colon N(f) \rightrightarrows A$, forming a pullback of f and f.

0.16 Classification of quotient objects A *quotient object* of an object A is represented by an epimorphism $e\colon A \to B$, and an epimorphism $e'\colon A \to B'$ represents the same quotient iff $e' = i \cdot e$ holds for some isomorphism $i\colon B \to B'$. We use the same adjective for quotient objects and (any of) the representing epimorphisms $e\colon A \to B$:

1. *Split* means that there exists $i\colon B \to A$ with $e \cdot i = \mathrm{id}_B$. Then B is called a *retract* of A.
2. *Regular* means that e is a coequalizer of a parallel pair with codomain A.
3. *Strong* means that in every commutative square

$$\begin{array}{ccc} A & \xrightarrow{e} & B \\ u\downarrow & \swarrow{d} & \downarrow v \\ X & \xrightarrow{m} & Z \end{array}$$

where m is a monomorphism, there is a "diagonal" morphism $d\colon B \to X$ such that $m \cdot d = v$ and $d \cdot e = u$.

4. *Extremal* means that in every commutative triangle

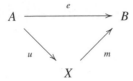

where m is a monomorphism, m is an isomorphism.

Dually, a *subobject* of A is represented by a monomorphism $m: B \to A$, and a monomorphism $m': B' \to A$ represents the same subobject iff $m' = m \cdot i$ holds for some isomorphism $i: B' \to B$.

0.17 Remark Let us recall some elementary facts on extremal, strong, and regular epimorphisms.

1. Every regular epimorphism is strong, and every strong epimorphism is extremal. If the category $\mathcal{A}$ has finite limits, then extremal = strong.
2. If the category $\mathcal{A}$ has binary products, then the condition of being an epimorphism in the definition of a strong epimorphism is redundant. The same holds for extremal epimorphisms if the category $\mathcal{A}$ has equalizers.
3. If a composite $f \cdot g$ is a strong epimorphism, then f is a strong epimorphism. The same holds for extremal epimorphisms, but in general, this fails for regular epimorphisms.
4. If f is a monomorphism and an extremal epimorphism, then it is an isomorphism.

0.18 Concrete categories Let $\mathcal{K}$ be a category.

1. By a *concrete category over* $\mathcal{K}$ is meant a category $\mathcal{A}$ together with a faithful functor $U: \mathcal{A} \to \mathcal{K}$.
2. Given concrete categories $(\mathcal{A}, U)$ and $(\mathcal{A}', U')$ over $\mathcal{K}$, a *concrete functor* is a functor $F: \mathcal{A} \to \mathcal{A}'$ such that $U = U' \cdot F$.

Further reading

For standard concepts of category theory, the reader may consult Adámek et al. (2009), Borceux (1994), or Mac Lane (1998).

1
Algebraic theories and algebraic categories

Algebras are classically presented by operations and equations. For example, the theory of groups is presented by three operations:

a binary operation $\circ$ (multiplication)
a unary operation $^{-1}$ (inverse)
a nullary operation e (unit)

and by the equations

$$x \circ (y \circ z) = (x \circ y) \circ z,$$
$$x \circ x^{-1} = e,$$
$$e \circ x = x,$$
$$x^{-1} \circ x = e,$$
$$x \circ e = x.$$

The whole equational theory of groups consists of all consequences of these five equations; for example, it contains the equation

$$x \circ (x^{-1} \circ x) = x.$$

The preceding presentation is not canonical: the first three equations are in fact sufficient, and so is the first one and the last two. There does not seem to exist any canonical minimal presentation of groups. But we can consider the equational theory as such. Following the idea of F. W. Lawvere from the 1960s, let us view the three basic group operations as maps:

$$\circ \colon G \times G \to G,$$
$$^{-1} \colon G \to G,$$
$$e \colon G^0 \to G.$$

10

All derived operations can be obtained via composition. For example, the operation $x \circ (y \circ z)$ is the composite

$$G \times (G \times G) \xrightarrow{\mathrm{id}_G \times \circ} G \times G \xrightarrow{\circ} G,$$

whereas the operation $(x \circ y) \circ z$ is the composite

$$(G \times G) \times G \xrightarrow{\circ \times \mathrm{id}_G} G \times G \xrightarrow{\circ} G.$$

The first equation in the above presentation is then an equality of these two compositions. The whole equational theory of groups is a category with finite products whose objects are integers $0, 1, 2, \ldots$ (representing the objects $G^0, G^1, G^2, \ldots$). The product corresponds to the addition of integers. Every n-ary operation correponds to a morphism from n to 1, while morphisms from n to m correspond to m-tuples of n-ary operations. For example, there is a morphism $G \times (G \times G) \to G \times G$ that corresponds to the pair of ternary operations

$$(x, y, z) \mapsto x$$
$$(x, y, z) \mapsto y \circ z.$$

Lawvere based his concept of an algebraic theory on these observations: algebraic theories are categories $\mathcal{T}$ with finite products whose objects are the natural numbers $0, 1, 2, \ldots$. Algebras are then functors from $\mathcal{T}$ to the category of sets that preserve finite products. Homomorphisms of algebras are represented by natural transformations.

We now introduce a more general definition of an algebraic theory and its algebras. See Chapter 11 for Lawvere's original concept of a one-sorted theory and Chapter 14 for S-sorted theories. In the present chapter, we also study basic concepts such as limits of algebras and representable algebras, and we introduce some of the main examples of algebraic categories.

1.1 Definition An *algebraic theory* is a small category $\mathcal{T}$ with finite products. An *algebra* for the theory $\mathcal{T}$ is a functor $A \colon \mathcal{T} \to \mathbf{Set}$ preserving finite products. We denote by $\mathrm{Alg}\,\mathcal{T}$ the category of algebras of $\mathcal{T}$. Morphisms, called *homomorphisms*, are the natural transformations; that is, $\mathrm{Alg}\,\mathcal{T}$ is a full subcategory of the functor category $\mathbf{Set}^{\mathcal{T}}$.

1.2 Definition A category is *algebraic* if it is equivalent to $\mathrm{Alg}\,\mathcal{T}$ for some algebraic theory $\mathcal{T}$.

1.3 Remark An algebraic theory is by definition a small category. However, throughout the book, we do not take care of the difference between *small* and *essentially small*: a category is essentially small if it is equivalent to a small category.

We will see in 10.15, 13.11, and 14.28 that algebraic categories correspond well to varieties, that is, equational categories of (many-sorted, finitary) algebras.

1.4 Example: *Sets* The simplest algebraic category is the category of sets itself. An algebraic theory $\mathcal{N}$ for *Set* can be described as the full subcategory of *Set*^{op} whose objects are the natural numbers $n = \{0, 1, \ldots, n-1\}$. In fact, since $n = 1 \times \ldots \times 1$ in *Set*^{op}, every algebra $A: \mathcal{N} \to$ *Set* is determined, up to isomorphism, by the set $A1$. More precisely, we have an equivalence functor

$$E: Alg\,\mathcal{N} \to Set, \quad A \mapsto A1.$$

The category *Set* has other algebraic theories – we describe them in Chapter 15, see Example 15.8.

1.5 Example: Many-sorted sets For a fixed set S, the power category *Set*^{S} of S-sorted sets and S-sorted functions is algebraic. A theory for *Set*^{S} can be described as the following category:

$$S^*,$$

whose objects are finite words over S (including the empty word). Morphisms from $s_0 \ldots s_{n-1}$ to $s'_0 \ldots s'_{k-1}$ are functions $a: k \to n$ such that $s_{a(i)} = s'_i$ ($i = 0, \ldots, k-1$). When S is the terminal set 1, this theory is nothing else than the theory $\mathcal{N}$ of *Set* described in 1.4. The fact that S^* is a theory of S-sorted sets will be seen in 1.18.

1.6 Example: abelian groups An algebraic theory for the category *Ab* of abelian groups is the category $\mathcal{T}_{ab}$ having natural numbers as objects, and morphisms from n to k are matrices of integers with n columns and k rows. The composition of $P: m \to n$ and $Q: n \to k$ is given by matrix multiplication $Q \cdot P = Q \times P: m \to k$, and identity morphisms are the unit matrices. If $n = 0$ or $k = 0$, the only $n \times k$ matrix is the empty one [ ]. $\mathcal{T}_{ab}$ has finite products. For example, 2 is the product 1×1 with projections $[1, 0]: 2 \to 1$ and $[0, 1]: 2 \to 1$. (In fact, given one-row matrices $P, Q: n \to 1$, there exists a unique two-row matrix $R: n \to 2$ such that $[1, 0] \cdot R = P$ and $[0, 1] \cdot R = Q$: the matrix with rows P and Q.) Here is a direct argument proving that the category *Ab* of abelian groups is equivalent to $Alg\,\mathcal{T}_{ab}$. Every abelian group G defines an algebra $\widehat{G}: \mathcal{T}_{ab} \to Set$ in the sense of 1.1 whose object function is

$\widehat{G}n = G^n$. For every morphism $P: n \to k$, we define $\widehat{G}P: G^n \to G^k$ by matrix multiplication:

$$\widehat{G}(P): \begin{bmatrix} g_1 \\ \vdots \\ g_n \end{bmatrix} \mapsto P \cdot \begin{bmatrix} g_1 \\ \vdots \\ g_n \end{bmatrix}.$$

The function $G \mapsto \widehat{G}$ extends to a functor $\widehat{(-)}: Ab \to Alg\, \mathcal{T}_{ab}$ in a rather obvious way: given a group homomorphism $h: G_1 \to G_2$, denote by $\hat{h}: \widehat{G}_1 \to \widehat{G}_2$ the natural transformation whose components are $h^n: G_1^n \to G_2^n$. It is clear that $\widehat{(-)}$ is a well-defined, full, and faithful functor. To prove that it is an equivalence functor, we need, for every algebra $A: \mathcal{T}_{ab} \to Set$, to present an abelian group G with $A \simeq \widehat{G}$. The underlying set of G is $A1$. The binary group operation is obtained from the morphism $[1, 1]: 2 \to 1$ in $\mathcal{T}_{ab}$ by $A[1, 1]: G^2 \to G$, the neutral element is $A[\,]: 1 \to G$ for the morphism $[\,]: 0 \to 1$ of $\mathcal{T}_{ab}$, and the operation of inverse is given by $A[-1]: G \to G$. It is not difficult to check that the axioms of abelian group are fulfilled. For example, the axiom $x + 0 = x$ follows from the fact that A preserves the composition of $\begin{bmatrix} 1 \\ 0 \end{bmatrix}: 1 \to 2$ with $[1, 1]: 2 \to 1$. Clearly $A \simeq \widehat{G}$ (consider the canonical isomorphism $An = A(1 \times \ldots \times 1) \simeq A1 \times \ldots \times A1 = G^n = \widehat{G}n$).

1.7 Example: Groups As mentioned at the beginning of this chapter, an algebraic theory of groups consists of all derived operations and all equations that hold in all groups. Viewed as an algebraic theory $\mathcal{T}_{gr}$, we have, as in the previous example, natural numbers as objects. Morphisms can be described by considering a standard set of variables $x_0, x_1, x_2, \ldots$ and defining the morphisms from n to 1 in $\mathcal{T}_{gr}$ to be precisely all terms in the variables $x_0, \ldots, x_{n-1}$ (see Remark 13.1 for a formal definition of the concept of a *term*). The morphisms from n to k in general are all k-tuples of morphisms in $\mathcal{T}_{gr}(n, 1)$. The identity morphism in $\mathcal{T}_{gr}(n, n)$ is the n-tuple of terms $(x_0, \ldots, x_{n-1})$, and composition is given by substitution of terms. We will see in Chapter 13 the reason why the category of groups is indeed equivalent to $Alg\, \mathcal{T}_{gr}$.

1.8 Example: Modules Let R be a ring with unit. The category $R\text{-}Mod$ of left modules and module homomorphisms is algebraic. A theory directly generalizing that of abelian groups has as objects natural numbers and as morphisms matrices over R. Algebraic categories of the form $R\text{-}Mod$ are treated in greater detail in Appendix B.

1.9 Example: One-sorted Σ-algebras Let Σ be a *signature*, that is, a set Σ (of operation symbols) together with an arity function

$$\text{ar}: \Sigma \to \mathbb{N}.$$

A Σ-*algebra* consists of a set A and, for every n-ary symbol $\sigma \in \Sigma$, an n-ary operation $\sigma^A: A^n \to A$. A homomorphism of Σ-algebras is a function preserving the given operations. The category

$$\Sigma\text{-}Alg$$

of Σ-algebras and homomorphisms (as well as its equational subcategories) are algebraic, as we demonstrate in Chapter 13. Clearly groups and abelian groups enter into this general example as well as many other classical algebraic structures such as Boolean algebras and Lie algebras.

1.10 Example: Many-sorted Σ-algebras Let S be a set (of sorts). An *S-sorted signature* is a set Σ (of operation symbols) together with an arity function

$$\text{ar}: \Sigma \to S^* \times S,$$

where S^* is the set of all finite words on S. We write $\sigma: s_1 \ldots s_n \to s$ for an operation symbol σ of arity $(s_1 \ldots s_n, s)$. In case $n = 0$, we write $\sigma: s$. (In 1.5, we have used the symbol S^* to denote a category, but there is no danger of confusion: it will always be clear from context whether we mean the category S^* or just the set of its objects.) A Σ-*algebra* consists of an S-sorted set $A = \langle A_s \rangle_{s \in S}$ and, for every symbol $\sigma \in \Sigma$ of arity $(s_1 \ldots s_n, s)$, an operation $\sigma^A: A_{s_1} \times \ldots \times A_{s_n} \to A_s$. Homomorphisms of Σ-algebras are S-sorted functions (i.e. morphisms of Set^S) preserving the given operations. The category

$$\Sigma\text{-}Alg$$

of Σ-algebras and homomorphisms is algebraic, and so are all equational subcategories, as we demonstrate in Chapter 14. Following is a concrete example.

1.11 Example: Graphs We denote by *Graph* the category of *directed graphs* G with multiple edges: they are given by a set G_v of vertices, a set G_e of edges, and two functions from G_e to G_v determining the target (τ) and the source (σ) of every edge. The morphisms are called *graph homomorphisms*: given graphs G and G', a graph homomorphism is a pair of functions $h_v: G_v \to G'_v$ and $h_e: G_e \to G'_e$ such that the source and the target of every edge are preserved. A theory for *Graph* will be described in 1.16.

1.12 Remark Every object t of an algebraic theory $\mathcal{T}$ yields the algebra $Y_{\mathcal{T}}(t)$, representable by t:

$$Y_{\mathcal{T}}(t) = \mathcal{T}(t, -) \colon \mathcal{T} \to Set.$$

Following 0.4, this and the Yoneda transformations define a full and faithful functor

$$Y_{\mathcal{T}} \colon \mathcal{T}^{op} \to Alg\,\mathcal{T}.$$

1.13 Lemma *For every algebraic theory $\mathcal{T}$, the Yoneda embedding*

$$Y_{\mathcal{T}} \colon \mathcal{T}^{op} \to Alg\,\mathcal{T}$$

preserves finite coproducts.

Proof If 1 is a terminal object of $\mathcal{T}$, then $\mathcal{T}(1, -)$ is an initial object of $Alg\,\mathcal{T}$: for every algebra A, we know that $A1$ is a terminal object, and thus there is a unique morphism $\mathcal{T}(1, -) \to A$.

Given two objects t_1, t_2 in $\mathcal{T}$, $\mathcal{T}(t_1 \times t_2, -)$ is a coproduct of $\mathcal{T}(t_1, -)$ and $\mathcal{T}(t_2, -)$ since for every algebra A, the morphisms $\mathcal{T}(t_1 \times t_2, -) \to A$ correspond to elements of $A(t_1 \times t_2) = A(t_1) \times A(t_2)$. $\square$

1.14 Example: Set-valued functors If $\mathcal{C}$ is a small category, the functor category $Set^{\mathcal{C}}$ is algebraic. An algebraic theory of $Set^{\mathcal{C}}$ is a *free completion* $\mathcal{T}_{\mathcal{C}}$ *of $\mathcal{C}$ under finite products*. This means that there exists a functor $E_{Th} \colon \mathcal{C} \to \mathcal{T}_{\mathcal{C}}$ such that $\mathcal{T}_{\mathcal{C}}$ is a category with finite products and for every functor $F \colon \mathcal{C} \to \mathcal{B}$, where $\mathcal{B}$ is a category with finite products, there exists an essentially unique functor (i.e., unique up to natural isomorphism) $F^* \colon \mathcal{T}_{\mathcal{C}} \to \mathcal{B}$, preserving finite products with F naturally isomorphic to $F^* \cdot E_{Th}$.

In other words, composition with E_{Th} gives an equivalence between the category of finite product–preserving functors from $\mathcal{T}_{\mathcal{C}}$ to $\mathcal{B}$ and the category of functors from $\mathcal{C}$ to $\mathcal{B}$. In particular, the categories $Set^{\mathcal{C}}$ and $Alg\,\mathcal{T}_{\mathcal{C}}$ are equivalent.

1.15 Remark

1. The free finite-product completion $\mathcal{T}_{\mathcal{C}}$ can be described as follows: objects of $\mathcal{T}_{\mathcal{C}}$ are all finite families

$$(C_i)_{i \in I}, \quad I \text{ finite}$$

of objects of $\mathcal{C}$, and morphisms from $(C_i)_{i \in I}$ to $(C'_j)_{j \in J}$ are pairs (a, α) where $a \colon J \to I$ is a function and $\alpha = (\alpha_j)_{j \in J}$ is a family of morphisms $\alpha_j \colon C_{a(j)} \to C'_j$ of $\mathcal{C}$. The composition and identity morphisms are defined as expected. A terminal object in $\mathcal{T}_\mathcal{C}$ is the empty family, and a product of two objects is the disjoint union of the families. Finally, the functor $E_{Th}\colon \mathcal{C} \to \mathcal{T}_\mathcal{C}$ is given by $E_{Th}(C) = (C)$. It is easy to verify the universal property: since for every object $(C_i)_{i \in I}$ in $\mathcal{T}_\mathcal{C}$, we have $(C_i)_{i \in I} = \Pi_I E_{Th}(C_i)$, then necessarily, $F^*((C_i)_{i \in I}) = \Pi_I F C_i$.

2. Equivalently, $\mathcal{T}_\mathcal{C}$ can be described as the category of all words over $\mathrm{obj}\,\mathcal{C}$ (the set of objects of $\mathcal{C}$); that is, objects have the form of n-tuples $C_0 \ldots C_{n-1}$, where each C_i is an object of $\mathcal{C}$ (and where n is identified with the set $\{0, \ldots, n-1\}$), including the case $n = 0$ (empty word). Morphisms from $C_0 \ldots C_{n-1}$ to $C'_0 \ldots C'_{k-1}$ are pairs (a, α) consisting of a function $a\colon k \to n$ and a k-tuple of $\mathcal{C}$-morphisms $\alpha = (\alpha_0, \ldots, \alpha_{k-1})$ with $\alpha_i\colon C_{a(i)} \to C'_i$.

1.16 Example The category of graphs (see 1.11) is equivalent to $Set^{\rightrightarrows}$, and its theory $\mathcal{T}_{graph}$ is the free completion of

$$\mathrm{id}_e \circlearrowright e \underset{\sigma}{\overset{\tau}{\rightrightarrows}} v \circlearrowleft \mathrm{id}_v$$

under finite products.

1.17 Remark Since the Yoneda embedding $Y_{\mathcal{T}_\mathcal{C}}\colon \mathcal{T}_\mathcal{C}^{op} \to Alg\,\mathcal{T}_\mathcal{C} \simeq Set^\mathcal{C}$ preserves finite coproducts (1.13), the category $\mathcal{T}_\mathcal{C}^{op}$ is equivalent to the full subcategory of $Set^\mathcal{C}$ given by finite coproducts of representable functors.

1.18 Example

1. The algebraic theory $\mathcal{N}$ for Set described in 1.4 is nothing else than the theory $\mathcal{T}_\mathcal{C}$ of 1.14 when $\mathcal{C}$ is the one-object discrete category.
2. More generally, if in 1.14 the category $\mathcal{C}$ is discrete, that is, it is a set S, $\mathcal{T}_\mathcal{C}$ is the theory S^* for S-sorted sets described in 1.5. Following 1.17, S^* is equivalent to the full subcategory of Set^S of *finite S-sorted sets* (an S-sorted set $\langle A_s \rangle_{s \in S}$ is finite if the coproduct $\coprod_S A_s$ is a finite set).

1.19 Example Another special case of Example 1.14 is the category M-Set of M-sets for a monoid M: if we consider M as a one-object category whose morphisms are the elements of M, then M-Set is equivalent to Set^M. As we will see in 13.15, M-Set is a category of unary algebras. More generally, the category $Set^\mathcal{C}$ can be presented as a category of unary S-sorted algebras: choose $S = \mathrm{obj}\,\mathcal{C}$ as set of sorts, choose $\Sigma = \mathrm{mor}\,\mathcal{C}$ as set of operation symbols, and define $\mathrm{ar}(f) = (s, s')$ if $f \in \mathcal{C}(s, s')$. Then $Set^\mathcal{C}$ is equivalent to the subcategory of Σ-Alg of those algebras satisfying the equations

$$u(v(x)) = (uv)(x) \quad (u, v \text{ composable morphisms of } \mathcal{C})$$

and

$$\mathrm{id}_s(x) = x \quad (s \in S).$$

1.20 Remark

1. In the example of abelian groups, we have the forgetful functor $U \colon Ab \to Set$ assigning to every abelian group its underlying set. Observe that the groups $\mathbb{Z}^n$ are free objects of Ab on n generators, and the full subcategory of all these objects is the dual of the theory $\mathcal{T}_{ab}$ in 1.6.
2. Analogously, if the category $\mathcal{C}$ of Example 1.14 has the object set S, then we have a forgetful functor that forgets the action of $A \colon \mathcal{C} \to Set$ on morphisms:

$$U \colon Set^{\mathcal{C}} \to Set^S, \quad UA = \langle A(s) \rangle_{s \in S}.$$

That functor U has a left adjoint

$$F \colon Set^S \to Set^{\mathcal{C}}, \quad F(\langle A_s \rangle_{s \in S}) = \coprod_{s \in S} \left(\coprod_{A_s} \mathcal{C}(s, -) \right)$$

(this easily follows from the Yoneda lemma because F preserves coproducts). Following 1.17, the objects of the theory $\mathcal{T}_{\mathcal{C}}$ are precisely the finitely generated free objects; this is the image of finite S-sorted sets under F.
3. In Chapters 11 and 14, we will see that this is not a coincidence: for every S-sorted algebraic category $\mathcal{A}$, the free objects on finite S-sorted sets form a full subcategory whose dual is a theory for $\mathcal{A}$ (see 11.22 for one-sorted algebraic categories and 14.13 for S-sorted algebraic categories).

The category of algebras of an algebraic theory is quite rich. We already know that every object t of an algebraic theory $\mathcal{T}$ yields the representable algebra $Y_{\mathcal{T}}(t) = \mathcal{T}(t, -)$. Other examples of algebras can be obtained, for example, by the formation of limits and colimits. We will now show that limits always exist and are built up at the level of sets. Also, colimits always exist, but they are seldom built up at the level of sets. We will study colimits in subsequent chapters.

1.21 Proposition *For every algebraic theory $\mathcal{T}$, the category $Alg\,\mathcal{T}$ is closed in $Set^{\mathcal{T}}$ under limits.*

Proof Limits are formed objectwise in $Set^{\mathcal{T}}$. Since limits and finite products commute (0.5), given a diagram in $Set^{\mathcal{T}}$ whose objects are functors preserving finite products, a limit of that diagram also preserves finite products. □

1.22 Corollary *Every algebraic category is complete.*

1.23 Remark

1. The previous proposition means that limits of algebras are formed objectwise at the level of sets. For example, a product of two graphs has both the vertex set given by the Cartesian product of the vertex sets and the set of edges given by the Cartesian product of the edge sets.
2. Monomorphisms in the category $Alg\,\mathcal{T}$ are precisely the homomorphisms that are componentwise monomorphisms (i.e., injective functions) in Set. In fact, this is true in $Set^{\mathcal{T}}$, and $Alg\,\mathcal{T}$ is closed under monomorphisms (being closed under limits) in $Set^{\mathcal{T}}$.
3. In every algebraic category, kernel pairs (0.15) exist and are formed objectwise (in Set).

1.24 Example One of the most important data types in computer science is a stack, or finite list, of elements of a set (of letters) called an *alphabet*. Here we consider stacks of natural numbers: we will have elements of sort n (a natural number) and s (a stack) and the following two-sorted signature:

succ: $n \to n$, the successor of a natural number
push: $sn \to s$, which adds a new letter to the leftmost position of a stack
pop: $s \to s$, which deletes the leftmost position
top: $s \to n$, which reads the top element of the stack

We will also have two constants: $e: s$, for the empty stack, and $0: n$. For simplicity, we put $\text{top}(e) = 0$ and $\text{pop}(e) = e$.

This leads us to the concept of algebras A of two sorts s (a stack) and n (a natural number) with operations

succ: $A_n \to A_n$,
push: $A_s \times A_n \to A_s$,
pop: $A_s \to A_s$,
top: $A_s \to A_n$

and with constants $0 \in A_n$ and $e \in A_s$.

We can consider stacks as equationally specified algebras of sorts $\{s, n\}$, and the algebraic theory is then obtained from the corresponding finitely generated free algebras.

1.25 Example: Sequential automata Recall that a deterministic sequential automaton A is given by a set A_s of states, a set A_i of input symbols, a set A_o of output symbols, and three functions:

$\delta \colon A_s \times A_i \to A_s$ (next-state function),

$\gamma \colon A_s \to A_o$ (output),

$\varphi \colon 1 \to A_s$ (initial state).

Given two sequential automata A and $A' = (A'_s, A'_i, A'_o, \delta', \gamma', \varphi')$, a morphism (simulation) is given by a triple of functions:

$$h_s \colon A_s \to A'_s, \quad h_i \colon A_i \to A'_i, \quad h_o \colon A_o \to A'_o,$$

such that the diagram

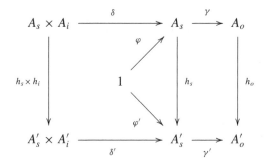

commutes. This is the category of algebras of three sorts s, i, and o, given by the signature

$$\delta \colon si \to s, \quad \gamma \colon s \to o, \quad \varphi \colon s.$$

Again, an algebraic theory of automata is formed by considering finitely generated free algebras.

Historical remarks

Algebraic theories were introduced by F. W. Lawvere in his dissertation (1963). He considered the one-sorted case, which we study in Chapter 11. This corresponds to (one-sorted) equational theories of Birkhoff (1935), which we treat in Chapter 13.

Many-sorted equational theories were first considered by Higgins (1963–1964) and were later popularized by Birkhoff and Lipson (1970). In a review of Higgins's paper, Heller (1965) suggested to look for the connection with

Lawvere's approach. This was done by Bénabou (1968), who dealt with many-sorted algebraic theories. Our definition of an algebraic theory is given without reference to sorting. This sort-free approach corresponds to the more general theory of sketches initiated by Ehresmann (1967) (see Bastiani & Ehresmann, 1972, for an exposition). The S-sorted approach is presented in Chapter 14.

The interested reader can find expositions of various aspects of algebraic theories in the following literature:

finitary theories and their algebras in general categories (Barr & Wells, 1985; Borceux, 1994; Hyland & Power, 2007; Pareigis, 1970; Pedicchio & Rovatti, 2004; Schubert, 1972).
infinitary theories (Linton, 1966; Wraith, 1970).
applications of theories in computer science (Barr & Wells, 1990; Wechler, 1992).

Manes (1976) is, in spite of its title, devoted to monads, not theories; an introduction to monads can be found in Appendix A.

2
Sifted and filtered colimits

Colimits in algebraic categories are, in general, not formed objectwise. In this chapter, we study the important case of sifted colimits, which are always formed objectwise. Prominent examples of sifted colimits are filtered colimits and reflexive coequalizers (see Chapter 3).

2.1 Definition A small category $\mathcal{D}$ is called

1. *sifted* if finite products in *Set* commute with colimits over $\mathcal{D}$
2. *filtered* if finite limits in *Set* commute with colimits over $\mathcal{D}$

Sifted (or filtered) diagrams are diagrams with a sifted (or filtered) domain. Colimits of sifted (or filtered) diagrams are called sifted (or filtered) colimits.

2.2 Remark

1. Explicitly, a small category $\mathcal{D}$ is sifted iff, given a diagram

$$D: \mathcal{D} \times \mathcal{J} \to Set,$$

   where $\mathcal{J}$ is a finite discrete category, the canonical map

$$\delta: \underset{\mathcal{D}}{\mathrm{colim}} \left(\prod_{\mathcal{J}} D(d, j) \right) \to \prod_{\mathcal{J}} (\underset{\mathcal{D}}{\mathrm{colim}}\, D(d, j)) \qquad (2.1)$$

   is an isomorphism. $\mathcal{D}$ is filtered iff it satisfies the same condition, but with respect to every finitely generated category $\mathcal{J}$, see 0.5 (replace $\prod_{\mathcal{J}}$ by $\lim_{\mathcal{J}}$ in (2.1)).
2. The canonical map δ in (2.1) is an isomorphism for every finite discrete category $\mathcal{J}$ iff δ is an isomorphism when $\mathcal{J}$ is the empty set and when $\mathcal{J}$ is the two-element set. The latter means that for every pair $D, D': \mathcal{D} \to Set$

of diagrams, the colimit of the diagram

$$D \times D' : \mathcal{D} \times \mathcal{D} \to Set, \quad (d, d') \mapsto Dd \times D'd'$$

is canonically isomorphic to *colim D* × *colim D'*.

2.3 Example

1. Colimits of ω-chains are filtered. Here the category $\mathcal{D}$ is the partially ordered set, or poset, of natural numbers, considered as a category.
2. More generally, colimits of chains (where $\mathcal{D}$ is an infinite ordinal considered as a poset) are filtered. These are the typical filtered colimits: a category having colimits of chains has all filtered colimits – see 1.5 and 1.7 in Adámek and Rosický (1994).
3. Generalizing still further: *directed colimits* are filtered. Recall that a poset is called (upward) directed if it is nonempty and every pair of elements has an upper bound. Directed colimits are colimits of diagrams whose schemes are directed posets.
4. An example of filtered colimits that are not directed are the colimits of idempotents. Let f be an endomorphism of an object A which is idempotent, that is, $f \cdot f = f$. This can be considered as a diagram whose domain $\mathcal{D}$ has one object and, besides the identity, precisely one idempotent morphism. This category is filtered. In fact, the colimit of the preceding diagram is the coequalizer of f and id_A. It is not difficult to verify directly (or using 2.19) that in *Set*, these coequalizers commute with finite limits. We return to colimits of idempotents in Chapter 8: they are precisely the splitting of idempotents studied there.

   There exist, essentially, no other finite filtered colimits than colimits of idempotents. In fact, whenever a finite category $\mathcal{D}$ is filtered, it has a cone $f_X : Z \to X$ ($X \in \mathrm{obj}\,\mathcal{D}$) over itself. It follows easily that $f_Z : Z \to Z$ is an idempotent, and a colimit of a diagram $D: \mathcal{D} \to \mathcal{A}$ exists iff the idempotent Df_Z has a colimit in $\mathcal{A}$.
5. Filtered colimits are, of course, sifted.
6. Coequalizers are colimits that are not sifted (see 2.17). As we will see in Chapter 3, reflexive coequalizers are sifted (but not filtered); these are coequalizers of parallel pairs $a_1, a_2 : A \rightrightarrows B$ for which $d: B \to A$ exists with $a_1 \cdot d = \mathrm{id}_B = a_2 \cdot d$.

   In fact, in a sense made precise in Chapter 7, we can state that

   $$\text{sifted colimits} = \text{filtered colimits} + \text{reflexive coequalizers}.$$

2.4 Remark Sifted categories have an easy characterization: they are nonempty and have, for every pair of objects, the category of all cospans connected.

This will be proved in 2.15. Before doing that, we need to recall the standard concepts of connected category and final functor. But we first present a result showing why sifted colimits are important.

2.5 Proposition *For every algebraic theory T, the category $Alg\, T$ is closed in Set^T under sifted colimits.*

Proof Since sifted colimits and finite products commute in *Set*, they do so in Set^T (where they are computed objectwise). It follows that a sifted colimit in Set^T of functors preserving finite products also preserves finite products. □

2.6 Example Coproducts are not sifted colimits. In fact, for almost no algebraic theory T is $Alg\, T$ closed under coproducts in Set^T. A concrete example: if T_{ab} is the theory of abelian groups (1.6), then binary coproducts in $Alg\, T_{ab}$ are products, and in $Set^{T_{ab}}$, they are disjoint unions.

2.7 Corollary *In every algebraic category, sifted colimits commute with finite products.*

In fact, this follows from the fact that the category $Alg\, T$ is closed under limits and sifted colimits in Set^T, and such limits and colimits in Set^T are formed objectwise.

2.8 Example In the category of abelian groups,

1. a directed union of abelian groups carries a canonical structure of an abelian group: this is the directed colimit of the diagram of inclusion homomorphisms and
2. let $a_1, a_2 \colon A \rightrightarrows B$ be a pair of homomorphisms with a common splitting $d \colon B \to A$ (i.e., $a_1 \cdot d = \mathrm{id}_B = a_2 \cdot d$) and let $c \colon B \to C$ be its coequalizer in *Set*; the set C carries a canonical structure of abelian group (the unique one for which c is a homomorphism).

   Reflexive coequalizers will be studied in detail in Chapter 3.

2.9 Remark Generalizing 2.8.1, a *directed union* in $Alg\, T$ is a directed colimit of subalgebras; that is, an algebra A is a directed union of subalgebras $m_i \colon A_i \to A$ ($i \in I$) provided that the poset on I given by $i \leq j$ iff $m_i \subseteq m_j$ is directed, and A is the colimit (with colimit cocone m_i) of the directed diagram of all

A_i, $i \in I$ and all connecting morphisms m_{ij}:

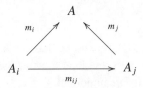

Thus $Alg\,\mathcal{T}$ is closed under directed unions in $Set^{\mathcal{T}}$.

2.10 Definition A category $\mathcal{A}$ is called *connected* if it is nonempty, and for every pair of objects X and X' in $\mathcal{A}$, there exists a zigzag of morphisms connecting X and X':

This is equivalent to saying that $\mathcal{A}$ cannot be decomposed as a coproduct (i.e., a disjoint union) of two nonempty subcategories.

2.11 Remark A small category $\mathcal{A}$ is connected iff the constant functor $\mathcal{A} \to Set$ of value 1 has colimit 1.

2.12 Definition A functor $F: \mathcal{D}' \to \mathcal{D}$ is called *final* if, for every diagram $D: \mathcal{D} \to \mathcal{A}$ such that $colim(D \cdot F)$ exists in $\mathcal{A}$, $colim\,D$ exists and the canonical morphism $colim\,D \cdot F \to colim\,D$ is an isomorphism.

2.13 Lemma *The following conditions on a functor $F: \mathcal{D}' \to \mathcal{D}$ are equivalent:*

1. *F is final.*
2. *F satisfies the finality condition with respect to all representable functors $D = \mathcal{D}(d, -)$.*
3. *For every object d of $\mathcal{D}$, the slice category $d \downarrow F$ of all morphisms $d \to Fd'$, $d' \in \operatorname{obj}\mathcal{D}'$, is connected.*

Proof The implication $1 \Rightarrow 2$ is trivial and $2 \Rightarrow 3$ follows from the usual description of colimits in Set (see 0.6) and the fact that since the diagram $\mathcal{D}(d, -)$ has colimit 1 (see 0.10), so does the diagram $\mathcal{D}(d, F-) = \mathcal{D}(d, -) \cdot F$.

To prove $3 \Rightarrow 1$, let $D: \mathcal{D} \to \mathcal{A}$ be a diagram and let $c_{d'}: D(Fd') \to C$ ($d' \in \operatorname{obj}\mathcal{D}'$) be a colimit of $D \cdot F$. For every object d of $\mathcal{D}$, choose a morphism $u_d: d \to Fd'$ for some $d' \in \operatorname{obj}\mathcal{D}'$ and put $g_d = c_{d'} \cdot Du_d: Dd \to C$.

We claim that $g_d: Dd \to C$ ($d \in \mathrm{obj}\,\mathcal{D}$) is a colimit of D. In fact, since $d \downarrow F$ is connected, it is easy to verify that g_d does not depend on the choice of d' and u_d and that these morphisms form a cocone of D. The rest of the proof is straightforward. □

2.14 Remark Following 2.13, the finality of the diagonal functor $\Delta: \mathcal{D} \to \mathcal{D} \times \mathcal{D}$ means that for every pair of objects A, B of $\mathcal{D}$, the category $(A, B) \downarrow \Delta$ of cospans on A, B is connected; that is;

1. a cospan $A \to X \leftarrow B$ exists
2. every pair of cospans on A, B is connected by a zigzag of cospans.

Therefore the statement

$\mathcal{D}$ is nonempty and the diagonal functor Δ is final

is equivalent to the statement

$\mathcal{D}$ is connected and the diagonal functor Δ is final.

2.15 Theorem *A small category $\mathcal{D}$ is sifted iff it is nonempty, and the diagonal functor $\Delta: \mathcal{D} \to \mathcal{D} \times \mathcal{D}$ is final.*

Proof We are going to prove that $\mathcal{D}$ is sifted iff it is connected and the diagonal functor Δ is final. More precisely, we are going to prove that

1. $\mathcal{D}$ is connected iff the canonical map δ (see 2.2) is an isomorphism when $\mathcal{J}$ is the empty set
2. Δ is final iff the canonical map δ is an isomorphism when $\mathcal{J}$ is the two-element set (i.e., is, binary products in *Set* commute with colimits over $\mathcal{D}$).

1. When $\mathcal{J}$ is the empty set, the codomain of δ is 1, whereas its domain is 1 iff $\mathcal{D}$ is connected (see 2.11).

2. Let $\mathcal{J}$ be the two-element set. Given diagrams $D, D': \mathcal{D} \to Set$, consider the functor

$$D \times D': \mathcal{D} \times \mathcal{D} \to Set, \quad (d, d') \mapsto Dd \times D'd'.$$

Since for every set X, the functor $X \times -: Set \to Set$ preserves colimits (see 0.11), the colimit of the diagram $D \times D'$ is

$$\operatorname*{colim}_{\mathcal{D} \times \mathcal{D}}(Dd \times D'd') \simeq \operatorname*{colim}_{d \in \mathcal{D}}(\operatorname*{colim}_{d' \in \mathcal{D}}(Dd \times D'd')) \simeq \operatorname*{colim}_{d \in \mathcal{D}} Dd \times \operatorname*{colim}_{d' \in \mathcal{D}} D'd'.$$

Consider now the following commutative diagram of canonical maps:

$$\begin{array}{ccc} \operatorname*{colim}_{\mathcal{D}}(Dd \times D'd) & \xrightarrow{\delta} & (\operatorname*{colim}_{\mathcal{D}} Dd) \times (\operatorname*{colim}_{\mathcal{D}} D'd) \\ = \downarrow & & \uparrow \simeq \\ \operatorname*{colim}_{\mathcal{D}}((D \times D') \cdot \Delta) & \xrightarrow{\beta} & \operatorname*{colim}_{\mathcal{D} \times \mathcal{D}}(D \times D') \end{array}$$

If Δ is final, then β is an isomorphism, and therefore δ is also an isomorphism.

Conversely, assume that δ is an isomorphism. Given two objects d and d' in $\mathcal{D}$, the representable functor $(\mathcal{D} \times \mathcal{D})((d, d'), -)$ is nothing but $D \times D'$: $\mathcal{D} \times \mathcal{D} \to \mathrm{Set}$, with $D = \mathcal{D}(d, -)$ and $D' = \mathcal{D}(d', -)$. If δ is an isomorphism, the previous diagram shows that Δ satisfies the finality condition with respect to all representable functors. Following 2.13, Δ is final. □

2.16 Example Every small category with finite coproducts is sifted. In fact, it contains an initial object, and the slice category $(A, B) \downarrow \Delta$ is connected because it has an initial object (the coproduct of A and B).

2.17 Example Consider the category $\mathcal{D}$ given by the morphisms

$$A \xrightarrow[g]{f} B$$

(identity morphisms are not depicted). $\mathcal{D}$ is not sifted. In fact, the slice category $(A, B) \downarrow \Delta$ is the discrete category with objects (f, id_B) and (g, id_B).

2.18 Remark Filtered colimits are closely related to sifted colimits. In fact, Definition 2.1 stresses this fact. The more usual definition of filtered category $\mathcal{D}$ is to say that every finitely generated subcategory of $\mathcal{D}$ has a cocone in $\mathcal{D}$ (this includes the condition that $\mathcal{D}$ is nonempty), and a well-known result states that this implies the property of Definition 2.1.2. The converse is also true, as follows.

2.19 Theorem *For a small category $\mathcal{D}$, the following conditions are equivalent:*

1. $\mathcal{D}$ *is filtered.*
2. *Every finitely generated subcategory of $\mathcal{D}$ has a cocone.*
3. $\mathcal{D}$ *is nonempty and fulfills the following:*
   a. *For every pair of object A, B, there exists a cospan $A \to X \leftarrow B$.*
   b. *For every parallel pair of morphisms $u, v \colon A \rightrightarrows B$, there exists a morphism $f \colon B \to C$ merging u and v: $f \cdot u = f \cdot v$.*

Proof The proof of the implications $3 \Leftrightarrow 2 \Rightarrow 1$ is standard; the reader can find it, for example, in Borceux (1994), volume 1, theorem 2.13.4. The proof of the implication $1 \Rightarrow 3$ is easy: for point a, argue as in 2.15; for point b use, analogously, the equalizer of $\mathcal{D}(u, -), \mathcal{D}(v, -) : \mathcal{D}(B, -) \rightrightarrows \mathcal{D}(A, -)$, which is the diagram D of all morphisms merging u and v: since $\operatorname{colim} D = 1$, the diagram is nonempty. □

With the next proposition, we establish a further analogy to 2.15.

2.20 Proposition *A small category $\mathcal{D}$ is filtered iff for any finitely generated category $\mathcal{J}$, see 0.5, the diagonal functor $\Delta \colon \mathcal{D} \to \mathcal{D}^{\mathcal{J}}$ is final.*

Proof 1. Let all such functors Δ be final. We are to show that every finitely generated subcategory $\mathcal{J}$ of $\mathcal{D}$ has a cocone in $\mathcal{D}$. The inclusion functor $d \colon \mathcal{J} \to \mathcal{D}$ is an object of the functor category $\mathcal{D}^{\mathcal{J}}$. Following 2.13.3, the slice category $d \downarrow \Delta$ is connected and thus nonempty. Since $d \downarrow \Delta$ is precisely the category of cocones of $\mathcal{J}$ in $\mathcal{D}$, we obtain the desired cocone.

2. Conversely, let $\mathcal{D}$ be filtered. We are to verify that for every object d of $\mathcal{D}^{\mathcal{J}}$, where $\mathcal{J}$ is finitely generated, the slice category $d \downarrow \Delta$ is connected.

2a. If $\mathcal{J}$ is a subcategory of $\mathcal{D}$ and $d \colon \mathcal{J} \to \mathcal{D}$ is the inclusion functor, then the category $d \downarrow \Delta$ of all cocones is nonempty. To prove that it is in fact connected, consider two cocones C_1 and C_2. Since $\mathcal{J}$ is finitely generated (say, by a finite set M of morphisms), it has finitely many objects; thus C_1 and C_2 are finite sets of morphisms. Put $\overline{M} = M \cup C_1 \cup C_2$ and let $\overline{\mathcal{J}}$ be the subcategory of $\mathcal{D}$ generated by $\overline{M}$. Then $\overline{\mathcal{J}}$ has a cocone in $\mathcal{D}$, and this cocone defines an obvious cocone of $\mathcal{J}$ with cocone morphisms to C_1 and C_2. Thus we obtain a zigzag of length 2.

2b. Let $d \colon \mathcal{J} \to \mathcal{D}$ be arbitrary and let M be a finite set of morphisms generating $\mathcal{J}$. Then the set

$$d(M) \cup \{\operatorname{id}_{d(x)} \,;\, x \in \operatorname{obj} \mathcal{J}\}$$

is finite and generates a subcategory $\mathcal{J}_0$ of $\mathcal{D}$. The slice category $d \downarrow \Delta$ is clearly equivalent to the category of cocones of $\mathcal{J}_0$ in $\mathcal{D}$, which is connected by point 2a. □

2.21 Remark Every colimit can be expressed as a filtered colimit of finite colimits; that is, given a diagram $D \colon \mathcal{D} \to \mathcal{A}$ with $\mathcal{A}$ cocomplete, $\operatorname{colim} D$ can be constructed as the filtered colimit of the diagram of all $\operatorname{colim} D'$, where $D' \colon \mathcal{D}' \to \mathcal{A}$ ranges over all domain restrictions of D to finitely generated subcategories $\mathcal{D}'$ of $\mathcal{D}$.

2.22 Remark In Chapter 7, we study functors preserving filtered and sifted colimits. In case of endofunctors of *Set*, these two properties coincide (see 6.30), but in general, the latter one is stronger (see 2.26). We use the following terminology.

2.23 Definition A functor is called *finitary* if it preserves filtered colimits.

2.24 Example Here we mention some endofunctors of *Set* that are finitary.

1. The functor
$$H_n\colon Set \to Set \quad H_n X = X^n$$
is finitary for every natural number n since finite products commute in *Set* with filtered colimits.
2. A coproduct of finitary functors is finitary.
3. Let Σ be a signature (1.9). We define the corresponding *polynomial functor*
$$H_\Sigma\colon Set \to Set$$
as the coproduct of the functors $H_{\mathrm{ar}(\sigma)}$ for $\sigma \in \Sigma$. Explicitly,
$$H_\Sigma X = \coprod_{n\in\mathbb{N}} \Sigma_n \times X^n,$$
where Σ_n is the set of all symbols of arity n ($n = 0, 1, 2, \ldots$). This is a finitary functor.

2.25 Example Let $H\colon Set \to Set$ be a functor. An *H-algebra* is a pair (A, a) where A is a set and $a\colon HA \to A$ is a function. A *homomorphism* from (A, a) to (B, b) is a function $f\colon A \to B$ such that the square

$$\begin{array}{ccc} HA & \xrightarrow{Hf} & HB \\ {\scriptstyle a}\downarrow & & \downarrow{\scriptstyle b} \\ A & \xrightarrow{f} & B \end{array}$$

commutes. The resulting category is denoted *H-Alg*.

1. In 13.23, we will see that if H is finitary, then the category *H-Alg* is algebraic.
2. The special case of a polynomial endofunctor H_Σ leads to Σ-algebras. Indeed, for every one-sorted signature Σ, the category Σ-*Alg* is precisely the category H_Σ-*Alg*: if (A, a) is a H_Σ-algebra, then the operations

σ^A: $A^n \to A$ are the domain restrictions of a to the summand A^n corresponding to $\sigma \in \Sigma_n$. This case will be treated in Chapter 13.

2.26 Example An example of a finitary functor not preserving sifted colimits is the forgetful functor U: *Pos* $\to$ *Set*, where *Pos* is the category of posets and order-preserving maps. Consider the reflexive pair u, v: $1 + 2 \rightrightarrows 2$ from the coproduct of the terminal poset 1 and the two-element chain, where both morphisms are identity on the second summand, and they map the first one to the top and bottom of 2, respectively. Whereas the coequalizer in *Pos* is given by the terminal poset, the coequalizer of Uu and Uv in *Set* has two elements.

Historical remarks

Filtered colimits are a natural generalization of directed colimits known from algebra and topology since the beginning of the twentieth century. The general concept can already be found in Bourbaki (1956), including the fact that directed colimits commute with finite products in *Set*. Both Artin et al. (1972) and Gabriel and Ulmer (1971) contain the general definition of filtered colimits and the fact that they are precisely those colimits that commute with finite limits in *Set*. Gabriel and Ulmer (1971) even speculated about the general commutation of colimits with limits in *Set* (see chapter 15) and characterized colimits commuting with finite products in *Set*; this is the source of Theorem 2.15. This was later rediscovered by Lair (1996), who called these colimits *tamisantes*. The term *sifted* was suggested by Peter Johnstone.

The concept of a final functor and the characterization Lemma 2.13 is a standard result of category theory, which can be found in Mac Lane (1998).

The fact that sifted colimits play an analogous role for algebraic categories as filtered colimits play for the locally finitely presentable categories was presented in Adámek and Rosický (2001).

3
Reflexive coequalizers

An important case of sifted colimits are reflexive coequalizers. We will see in Chapter 7 that in algebraic categories, sifted colimits are just a combination of filtered colimits and reflexive coequalizers. A special case of reflexive coequalizers are coequalizers of equivalence relations that correspond to the classical concept of congruence. It turns out that the step from those classical coequalizers to all reflexive coequalizers makes the theory clearer from a categorical perspective.

3.1 Definition *Reflexive coequalizers* are coequalizers of reflexive pairs, that is, parallel pairs of split epimorphisms having a common splitting.

3.2 Remark In other words, reflexive coequalizers are colimits of diagrams over the category $\mathcal{M}$ given by the morphisms

$$A \underset{a_2}{\overset{a_1}{\rightleftarrows}} \overset{d}{\leftarrow} B$$

(identity morphisms are not depicted) composed freely modulo $a_1 \cdot d = \mathrm{id}_B = a_2 \cdot d$. This category is sifted: it is an easy exercise to check that the categories $(A, A) \downarrow \Delta$, $(A, B) \downarrow \Delta$ and $(B, B) \downarrow \Delta$ are connected.

Another method of verifying that $\mathcal{M}$ is a sifted category is to prove directly that reflexive coequalizers commute in *Set* with binary products. In fact, suppose that

$$A \underset{a_2}{\overset{a_1}{\rightrightarrows}} B \overset{c}{\longrightarrow} C \quad \text{and} \quad A' \underset{a'_2}{\overset{a'_1}{\rightrightarrows}} B' \overset{c'}{\longrightarrow} C'$$

are reflexive coequalizers in *Set*. We can assume, without loss of generality, that c is the canonical function of the quotient $C = B/\sim$ modulo the equivalence relation described as follows: two elements $x, y \in B$ are equivalent iff there exists a zigzag

where $i_1, i_2, \ldots, i_{2k}$ are 1 or 2. For reflexive pairs a_1, a_2, the zigzags can always be chosen to have the form

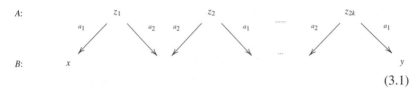
(3.1)

(here, for the elements z_{2i} of A, we use a_1, a_2, and for the elements z_{2i+1}, we use a_2, a_1). In fact, let $d: B \to A$ be a joint splitting of a_1, a_2. Given a zigzag, say,

we can modify it as follows:

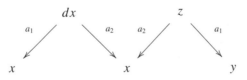

and analogously for the general case. Moreover, the length $2k$ of the zigzag (3.1) can be prolonged to $2k + 2$ or $2k + 4$ etc. by using d. Analogously, we can assume $C' = B'/\sim'$, where $\sim'$ is the equivalence relation given by zigzags of a'_1 and a'_2 of the preceding form (3.1). Now we form the parallel pair

$$A \times A' \xrightarrow[a_2 \times a'_2]{a_1 \times a'_1} B \times B'$$

and obtain its coequalizer by the zigzag equivalence $\approx$ on $B \times B'$. Given $(x, x') \approx (y, y')$ in $B \times B'$, we obviously have zigzags both for $x \sim y$ and for $x' \sim' y'$ (use projections of the given zig-zag), but also the other way around: whenever $x \sim y$ and $x' \sim' y'$, we choose the two zigzags so that they both have the preceding type (3.1) and the same lengths. They create an obvious zig-zag for $(x, x') \approx (y, y')$. From this it follows that the map

$$A \times A' \xrightarrow[a_2 \times a'_2]{a_1 \times a'_1} B \times B' \xrightarrow{c \times c'} (B/\sim) \times (B'/\sim')$$

is a coequalizer, as required.

3.3 Corollary *For every algebraic theory $\mathcal{T}$, the category $\operatorname{Alg} \mathcal{T}$ is closed in $\operatorname{Set}^{\mathcal{T}}$ under reflexive coequalizers.*

In fact, this follows from 2.5 and Remark 3.2.

3.4 Example In a category with kernel pairs, every regular epimorphism is a reflexive coequalizer. In fact, if r_1, r_2 is a kernel pair of a regular epimorphism $e \colon A \to B$,

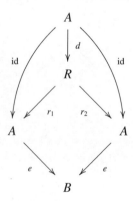

then e is a coequalizer of r_1, r_2. And since $e \cdot \operatorname{id} = e \cdot \operatorname{id}$, there exists a unique d with $r_1 \cdot d = \operatorname{id} = r_2 \cdot d$.

3.5 Corollary *For every algebraic theory $\mathcal{T}$, the category $\operatorname{Alg} \mathcal{T}$ is closed in $\operatorname{Set}^{\mathcal{T}}$ under regular epimorphisms. Therefore regular epimorphisms in $\operatorname{Alg} \mathcal{T}$ are precisely the homomorphisms that are componentwise epimorphisms (i.e., surjective functions) in Set.*

In fact, the first part of the statement follows from 2.5 and Example 3.4. The second statement is clear since the property holds in $\operatorname{Set}^{\mathcal{T}}$.

3.6 Remark In particular, every algebraic category is *co-wellpowered* with respect to regular epimorphisms. This means that for a fixed object A, the regular quotients of A constitute a set (not a proper class). In fact, this is true in *Set* and therefore, by Corollary 3.5, in every algebraic category.

More is true: algebraic categories are co-wellpowered with respect to all epimorphisms. We do not present a proof of this fact here because we do not need it. The interested reader can consult Adámek and Rosický (1994), sections 1.52 and 1.58.

In classical algebra, the homomorphism theorem states that every homomorphism can be factorized as a surjective homomorphism followed by the inclusion of a subalgebra. This holds in general algebraic categories; as follows.

3.7 Corollary *Every algebraic category has* regular factorizations, *that is, every morphism is a composite of a regular epimorphism followed by a monomorphism.*

Proof The category Set^T has regular factorizations: given a morphism $f\colon A \to B$, form a kernel pair $r_1, r_2\colon R \rightrightarrows A$ and its coequalizer $e\colon A \to C$. The factorizing morphism m,

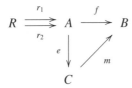

is a monomorphism. This follows from the fact that kernel pairs and coequalizers are formed objectwise (in *Set*). Since $Alg\,T$ is closed in Set^T under kernel pairs (1.23) and their coequalizers (3.3), it inherits the regular factorizations from Set^T. □

3.8 Example In *Ab* we know that

1. coproducts are not formed at the level of sets; in fact, $A + B = A \times B$ for all abelian groups A, B, and
2. reflexive coequalizers are formed at the level of sets, but general coequalizers are not; consider, for example, the pair $x \mapsto 2x$ and $x \mapsto 0$ of endomorphisms of $\mathbb{Z}$ whose coequalizer in *Ab* is finite and in *Set* is infinite.

3.9 Remark We provided a simple characterization of monomorphisms (1.23) and regular epimorphisms (3.5) in algebraic categories. There does not seem to be a simple characterization of the dual concepts (epimorphisms and regular monomorphisms). In fact, there exist algebraic categories with nonsurjective epimorphisms and with nonregular monomorphisms, as we show in the following example.

3.10 Example: Monoids These are algebras with one associative binary operation and one constant that is a neutral element. The category *Mon* of monoids and homomorphisms is algebraic (see Example 13.14).

An example of an epimorphism that is not regular is the embedding

$$i\colon \mathbb{Z} \to \mathbb{Q}$$

of the multiplicative monoid of integers into that of rational numbers. In fact, consider monoid homomorphisms $h, k \colon \mathbb{Q} \to A$ such that $h \cdot i = k \cdot i$; that is, $h(n) = k(n)$ for every integer n. To prove $h = k$, it is sufficient to verify $h(1/m) = k(1/m)$ for all integers $m \neq 0$: this follows from $h(m) \cdot h(1/m) = k(m) \cdot k(1/m) = 1$ (since $h(1) = k(1) = 1$). Consequently, i is not a regular epimorphism. Observe that i is also a monomorphism but not a regular one.

3.11 Remark Recall that in a finitely complete category $\mathcal{A}$, *relations* on an object A are the subobjects of $A \times A$. A relation can be represented by a monomorphism $r \colon R \to A \times A$ or by a parallel pair $r_1, r_2 \colon R \rightrightarrows A$ of morphisms that are jointly monic. The following definitions generalize the corresponding concepts for relations in *Set*.

3.12 Definition A relation $r_1, r_2 \colon R \rightrightarrows A$ in a category $\mathcal{A}$ is called

1. *reflexive* if the pair r_1, r_2 is reflexive ($r_1 \cdot d = \mathrm{id} = r_2 \cdot d$ for some $d \colon A \to R$)
2. *symmetric* if there exists $s \colon R \to R$ with $r_1 = r_2 \cdot s$ and $r_2 = r_1 \cdot s$
3. *transitive*, provided that for a pullback $\overline{R}$ of r_1 and r_2, there exists a morphism $t \colon \overline{R} \to R$ such that the diagram

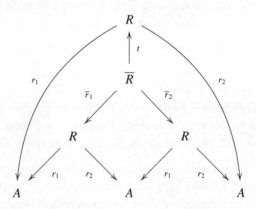

commutes.

An *equivalence relation* is a relation that is reflexive, symmetric, and transitive.

3.13 Remark

1. If $r_1, r_2 \colon R \rightrightarrows A$ is an equivalence relation, then the morphisms d, s and t of Definition 3.12 are necessarily unique.
2. An equivalence relation in *Set* is precisely an equivalence relation in the usual sense.
3. Given a relation $r_1, r_2 \colon R \rightrightarrows A$ and an object X, we can define a relation $\sim_R$ on the hom-set $\mathcal{A}(X, A)$ as follows: $f \sim_R g$ if there exists a morphism $h \colon X \to R$ such that $r_1 \cdot h = f$ and $r_2 \cdot h = g$. It is easy to check that $r_1, r_2 \colon R \rightrightarrows A$ is an equivalence relation in $\mathcal{A}$ iff $\sim_R$ is an equivalence relation in *Set* for all X in $\mathcal{A}$.
4. Kernel pairs are equivalence relations.
5. In the category of Σ-algebras (see 1.9), an equivalence relation on an algebra A is precisely a subobject of $A \times A$, which, as a relation on the underlying set of A, is an equivalence relation in *Set*. These relations are usually called congruences on A. We refer to them as equivalence relations in Σ-*Alg* because the concept of *congruence* is reserved (with the exceptions only of 11.31–11.33) for congruences of algebraic theories (see Chapter 10).

3.14 Definition A category is said to have *effective equivalence relations* provided that every equivalence relation is a kernel pair.

3.15 Example 1. All algebraic categories have effective equivalence relations, see Corollary 3.18.

2. The category of posets does not have this property: take an arbitrary poset B and an equivalence relation R on the underlying set of B equipped with the discrete ordering; the two projections $R \rightrightarrows B$ form an equivalence relation that is seldom a kernel pair.

3.16 Definition A category is called *exact* if it has

1. finite limits
2. coequalizers of kernel pairs
3. effective equivalence relations and
4. regular epimorphisms stable under pullback; that is, in every pullback

if e is a regular epimorphism, then so is e'.

3.17 Example *Set* is an exact category. In fact,

1. if $r_1, r_2 \colon R \rightrightarrows A$ is an equivalence relation, its coequalizer is

$$q \colon A \to A/\sim_R$$

where $\sim_R$ is as in Remark 3.13 (with $X = 1$) and q is the canonical morphism. Clearly $r_1, r_2 \colon R \rightrightarrows A$ is a kernel pair of q.

2. Given a pullback as in Definition 3.16 and an element $x \in B$, we choose $z \in C$ with $g(x) = e(z)$ using the fact that e is an epimorphism. Then (x, z) is an element of the pullback A and $e'(x, z) = x$. This proves that e' is surjective.

3.18 Corollary *Every algebraic category is exact.*

In fact, since *Set* is exact, so is $Set^{\mathcal{T}}$. Since equivalence relations are reflexive pairs, the exactness of $Alg\,\mathcal{T}$ follows using 1.23 and Corollary 3.3.

3.19 Definition We say that colimits in a category $\mathcal{A}$ *distribute over products* if given diagrams $D_i \colon \mathcal{D}_i \to \mathcal{A}$ ($i \in I$), and forming the diagram

$$D \colon \prod_{i \in I} \mathcal{D}_i \to \mathcal{A}, \quad Dd_i = \prod_{i \in I} D_i d_i,$$

the canonical morphism

$$\operatorname{colim} D \to \prod_{i \in I} \operatorname{colim} D_i$$

is an isomorphism.

If all $\mathcal{D}_i$ are of a certain type, we say that colimits of that type distribute over products. The concept of *distributing over finite products* is defined analogously but I is required to be finite.

3.20 Example In the category *Set*, it is easy to verify that

1. filtered colimits distribute over products
2. all colimits distribute over finite products

However, reflexive coequalizers do not distribute over infinite products. In fact, consider the coequalizers

$$n + n \underset{g_n}{\overset{f_n}{\rightrightarrows}} n + 1 \xrightarrow{c_n} 1,$$

where the left-hand components of f_n and g_n are the inclusion maps $n \to n + 1$ and the right-hand ones are $i \mapsto i$ and $i \mapsto i + 1$, respectively. Then $\prod_{n \in \mathbb{N}} f_n, \prod_{n \in \mathbb{N}} g_n$ have a coequalizer with infinite codomain and thus the coequalizer is distinct from $\prod_{n \in \mathbb{N}} c_n$.

3.21 Corollary *In every algebraic category,*

1. *regular epimorphisms are stable under products: given regular epimorphisms $e_i: A_i \to B_i$ ($i \in I$),*

$$\prod_{i \in I} e_i : \prod_{i \in I} A_i \to \prod_{i \in I} B_i$$

   is a regular epimorphism
2. *filtered colimits distribute over products*
3. *sifted colimits distribute over finite products*

In fact, since each of the three statements holds in *Set*, they hold in $Set^{\mathcal{T}}$, where limits and colimits are formed objectwise. Following Propositions 1.21 and 2.5, and Corollary 3.5, the statements hold in $Alg\,\mathcal{T}$ for every algebraic theory $\mathcal{T}$.

3.22 Remark

1. In the preceding corollary, we implicitly assume the existence of colimits in an algebraic category. This is not a restriction because every algebraic category is cocomplete, as we prove in the next chapter.
2. Although in *Set*, all colimits distribute over finite products, this is not true in algebraic categories in general: consider the empty diagram in the category of unitary rings. For $I = \{1, 2\}$ in Definition 3.19 and $\mathcal{D}_1 = \emptyset = \mathcal{D}_2$, we get *colim* $D = \mathbb{Z}$ and *colim* $D_1 \times$ *colim* $D_2 = \mathbb{Z} \times \mathbb{Z}$.

Historical remarks

Reflexive coequalizers were probably first applied by Linton (1969a). That they commute with finite products in *Set* (in fact, in every Cartesian closed category) is contained in the unpublished thesis of Johnstone written in the early 1970s. The importance of reflexive coequalizers for algebraic categories was first understood by Diers (1976), but his paper remained unnoticed. It was later rediscovered by Pedicchio and Wood (2000). A decisive step was taken in Adámek et al. (2001a, 2003), where reflexive coequalizers were used for establishing the algebraic duality (see Chapter 9) and for the study of abstract operations, different from limits and filtered colimits, performed in algebraic categories.

Effective equivalence relations were introduced by Artin et al. (1972) and exact categories by Barr et al. (1971). The fact that filtered colimits distribute with all products in *Set* goes back to Artin et al. (1972).

4
Algebraic categories as free completions

In this chapter, we prove that every algebraic category has colimits. Moreover, the category $Alg\,\mathcal{T}$ is a free completion of $\mathcal{T}^{op}$ under sifted colimits. This shows that algebraic categories can be characterized by their universal property: they are precisely the free sifted-colimit completions of small categories with finite coproducts. This is analogous to the classical result of Gabriel and Ulmer characterizing locally finitely presentable categories as precisely the categories $Ind\,\mathcal{C}$, where $\mathcal{C}$ is a small category with finite colimits and Ind is the free completion under filtered colimits (see Example 4.12).

4.1 Remark For the existence of colimits, since we already know that $Alg\,\mathcal{T}$ has sifted colimits and, in particular, reflexive coequalizers (see 2.5 and 3.3), all we need to establish is the existence of finite coproducts. Indeed, coproducts then exist because they are filtered colimits of finite coproducts. And coproducts and reflexive coequalizers construct all colimits (see 0.7). The first step toward the existence of finite coproducts has already been done in Lemma 1.13: finite coproducts of representable algebras, including an initial object, exist in $Alg\,\mathcal{T}$. In the next lemma, we use the category of elements $El\,A$ of a functor $A: \mathcal{T} \to Set$ introduced in 0.14.

4.2 Lemma *Given an algebraic theory $\mathcal{T}$, for every functor A in $Set^{\mathcal{T}}$, the following conditions are equivalent:*

1. *A is an algebra.*
2. *$El\,A$ is a sifted category.*
3. *A is a sifted colimit of representable algebras.*

Proof $2 \Rightarrow 3$: This follows from 0.14.

$3 \Rightarrow 1$: Representable functors are objects of $Alg\,T$ (1.12), and $Alg\,T$ is closed in Set^T under sifted colimits (2.5).

$1 \Rightarrow 2$: Following 2.16, it suffices to prove that $(El\,A)^{op}$ has finite products. This is obvious: for example, the product of (X, x) and (Z, z) is $(X \times Z, (x, z))$ – recall that $(x, z) \in AX \times AZ = A(X \times Z)$. $\square$

4.3 Remark An analogous result (with a completely analogous proof) holds for small categories T with finite limits: a functor $A\colon T \to Set$ preserves finite limits iff $El\,A$ is a filtered category iff A is a filtered colimit of representable functors.

4.4 Lemma

1. *If two functors $F\colon \mathcal{D} \to \mathcal{C}$ and $G\colon \mathcal{B} \to \mathcal{A}$ are final, then the product functor $F \times G\colon \mathcal{D} \times \mathcal{B} \to \mathcal{C} \times \mathcal{A}$ is final.*
2. *A product of two sifted categories is sifted.*

Proof
1. This follows from 2.13.3 because for any object (c, a) in $\mathcal{C} \times \mathcal{A}$, the slice category $(c, a) \downarrow F \times G$ is nothing but the product category $(c \downarrow F) \times (a \downarrow G)$, and the product of two connected categories is connected.

2. It is obvious from the preceding and 2.15. $\square$

4.5 Theorem *Every algebraic category is cocomplete.*

Proof As explained at the beginning of this chapter, we only need to establish finite coproducts $A + B$ in $Alg\,T$. Express A as a sifted colimit of representable algebras (Lemma 4.2)

$$A = colim\,(Y_T \cdot \Phi_A)$$

and analogously for B. The category

$$\mathcal{D} = El\,A \times El\,B$$

is sifted by Lemma 4.4, and for the projections P_1, P_2 of $\mathcal{D}$, we have two colimits in $Alg\,T$ over $\mathcal{D}$:

$$A = colim\,Y_T \cdot \Phi_A \cdot P_1 \quad \text{and} \quad B = colim\,Y_T \cdot \Phi_B \cdot P_2.$$

The diagram $D\colon \mathcal{D} \to Alg\,T$ assigning to every pair (X, x) and (Z, z) a coproduct of the representable algebras (see 1.13)

$$D((X, x), (Z, z)) = Y_T \cdot \Phi_A(x) + Y_T \cdot \Phi_B(z) \quad (\text{in } Alg\,T)$$

is sifted, thus it has a colimit in $Alg\ \mathcal{T}$. Since colimits over $\mathcal{D}$ commute with finite coproducts, we get

$$colim\ D = \underset{(x,z)}{colim}\ Y_\mathcal{T} \cdot \Phi_A(x) + \underset{(x,z)}{colim}\ Y_\mathcal{T} \cdot \Phi_B(z) = A + B.$$

$\square$

4.6 Example: Coproducts

1. In the category Ab of abelian groups, finite coproducts are finite products: the abelian group $A \times B$ together with the homomorphisms

$$\langle id_A, 0 \rangle \colon A \to A \times B \quad \text{and} \quad \langle 0, id_B \rangle \colon B \to A \times B$$

   is a coproduct of A and B.
2. Infinite coproducts $\coprod_{i \in I} A_i$ are directed colimits of finite subcoproducts $\coprod_{j \in J} A_j = \prod_{j \in J} A_j$ (for $J \subseteq I$ finite).
3. In the category of sequential automata (1.25), the product $A \times B$ of two automata is the machine working simultaneously in A and B on the given (joint) input streams, whereas the coproduct $A + B$ is the machine working, on a given input stream, completely in A or completely in B.
4. A coproduct of graphs in $Graph$ is given by the disjoint union of vertices and the disjoint union of edges.

4.7 Example: Coequalizers

1. In Ab, a coequalizer of homomorphisms $f, g \colon A \rightrightarrows B$ is the quotient $c \colon B \to B/B_0$ modulo the subgroup $B_0 \subseteq B$ of the elements $f(a) - g(a)$ for all $a \in A$.
2. A coequalizer of a parallel pair $f, g \colon A \rightrightarrows B$ in $Graph$ is given by forming the coequalizers in Set of (1) the two vertex functions and (2) the two edge functions.

4.8 Remark Before characterizing algebraic categories as free completions under sifted colimits, let us recall the general concept of a free completion of a category $\mathcal{C}$: this is, roughly speaking, a cocomplete category $\mathcal{A}$ in which $\mathcal{C}$ is a full subcategory such that every functor from $\mathcal{C}$ to a cocomplete category has an *essentially unique* extension (i.e., unique up to natural isomorphism) to a colimit-preserving functor with domain $\mathcal{A}$. In the following definition, we say this more precisely. Also, for a given class $\mathbb{D}$ of small categories, we define a free completion under $\mathbb{D}$-colimits, meaning that all colimits considered are colimits of diagrams with domains that are elements of $\mathbb{D}$.

4.9 Definition Let $\mathbb{D}$ be a class of small categories. By a *free completion of a category $\mathcal{C}$ under $\mathbb{D}$-colimits* is meant a functor $E_\mathbb{D} \colon \mathcal{C} \to \mathbb{D}(\mathcal{C})$ such that

Algebraic categories as free completions

1. $\mathbb{D}(\mathcal{C})$ is a category with $\mathbb{D}$-colimits
2. for every functor $F: \mathcal{C} \to \mathcal{B}$, where $\mathcal{B}$ is a category with $\mathbb{D}$-colimits, there exists an essentially unique functor $F^*: \mathbb{D}(\mathcal{C}) \to \mathcal{B}$ preserving $\mathbb{D}$-colimits with F naturally isomorphic to $F^* \cdot E_\mathbb{D}$:

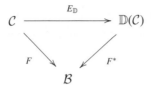

If $\mathbb{D}$ consists of all small categories, then $E_\mathbb{D}: \mathcal{C} \to \mathbb{D}(\mathcal{C})$ is called a *free completion of $\mathcal{C}$ under colimits*.

4.10 Theorem *For every small category $\mathcal{C}$, the Yoneda embedding*

$$Y_{\mathcal{C}^{op}}: \mathcal{C} \to Set^{\mathcal{C}^{op}}$$

is a free completion of $\mathcal{C}$ under colimits.

Proof The category $Set^{\mathcal{C}^{op}}$ is of course cocomplete. Let $F: \mathcal{C} \to \mathcal{B}$ be a functor, where $\mathcal{B}$ has colimits. Since $F^*: Set^{\mathcal{C}^{op}} \to \mathcal{B}$ should extend F and preserve colimits, we are forced to define it on objects $A: \mathcal{C}^{op} \to Set$ (using the notation of 0.14 applied to $\mathcal{T} = \mathcal{C}^{op}$) by

$$F^*A = \underset{El\,A}{colim}(F \cdot \Phi_A).$$

The definition on morphisms (i.e., natural transformations) $h: A_1 \to A_2$ is also obvious: h induces a functor $El\,h: El\,A_1 \to El\,A_2$, which to every element (X, x) of A_1 assigns the corresponding element $(X, h_X(x))$ of A_2. By the universal property of colimits, $El\,h$ induces a morphism

$$h': colim(F \cdot \Phi_{A_1}) \to colim(F \cdot \Phi_{A_2}),$$

and we are forced to define $F^*h = h'$.

The rule $A \mapsto colim(F \cdot \Phi_A)$ above defines a functor $F^*: Set^{\mathcal{C}^{op}} \to \mathcal{B}$, which fulfills $F^* \cdot Y_{\mathcal{C}^{op}} \simeq F$ because for $A = Y_{\mathcal{C}^{op}}(X) = \mathcal{C}(-, X)$, a colimit of $F \cdot \Phi_A = \mathcal{B}(-, FX)$ is FX. It remains to prove that F^* preserves colimits: for this we prove that F^* has the following right adjoint:

$$R: \mathcal{B} \to Set^{\mathcal{C}^{op}}, \quad RB = \mathcal{B}(F-, B).$$

We prove the adjunction $F^* \dashv R$ by verifying that there is a bijection

$$\mathcal{B}(F^*A, B) \simeq Set^{\mathcal{C}^{op}}(A, RB)$$

natural in $A\colon \mathcal{C}^{op} \to Set$ and $B \in \mathcal{B}$. In fact, the definition of F^* makes it clear that the left-hand side consists of precisely all cocones of the diagram $F \cdot \Phi_A$ with codomain B in $\mathcal{B}$:

$$\mathcal{B}(F^*A, B) = \mathcal{B}(colim(F \cdot \Phi_A), B) \simeq lim\, \mathcal{B}(F(\Phi_A(X, x)), B).$$

The same is true about the right-hand side: recall that $A = colim(Y_{\mathcal{C}^{op}} \cdot \Phi_A)$ (0.14), thus a natural transformation from A to RB is a cocone of the diagram $Y_{\mathcal{C}^{op}} \cdot \Phi_A$ with codomain RB:

$$Set^{\mathcal{C}^{op}}(A, RB) = Set^{\mathcal{C}^{op}}(colim(Y_{\mathcal{C}^{op}} \cdot \Phi_A), RB)$$
$$\simeq lim\, Set^{\mathcal{C}^{op}}(Y_{\mathcal{C}^{op}}(\Phi_A(X, x)), RB).$$

The Yoneda lemma tells us that morphisms from the objects $Y_{\mathcal{C}^{op}}(X)$ of that diagram to $RB = \mathcal{B}(F-, B)$ are precisely the members of the set $\mathcal{B}(FX, B)$:

$$Set^{\mathcal{C}^{op}}(Y_{\mathcal{C}^{op}}(X), RB) \simeq \mathcal{B}(FX, B).$$

In this sense, morphisms from A to RB in $Set^{\mathcal{C}^{op}}$ encode precisely the cocones of $F \cdot \Phi_A$ with codomain B. $\square$

4.11 Remark

1. Although the triangle in Definition 4.9 commutes up to natural isomorphism only, in Theorem 4.10 it is actually always possible to choose F^* so that the (strict) equality

$$F = F^* \cdot Y_{\mathcal{C}}$$

holds. This is easily seen from the preceding proof since if the algebra A has the form $A = Y_{\mathcal{C}^{op}}(X)$, a colimit of $F \cdot \Phi_A$ can be chosen to be FX.
2. Let $Colim(Set^{\mathcal{C}^{op}}, \mathcal{B})$ be the full subcategory of $\mathcal{B}^{Set^{\mathcal{C}^{op}}}$ of all functors preserving colimits. Then composition with $Y_{\mathcal{C}^{op}}$ defines a functor

$$- \cdot Y_{\mathcal{C}^{op}}\colon Colim(Set^{\mathcal{C}^{op}}, \mathcal{B}) \to \mathcal{B}^{\mathcal{C}}.$$

The preceding universal property tells us that this functor is an equivalence. (It is, however, not an isomorphism of categories, even assuming the choice of F^* in point 1.)

4.12 Example

1. A famous classical example is the free completion under filtered colimits denoted by

$$E_{Ind}\colon \mathcal{C} \to Ind\,\mathcal{C}.$$

For a small category $\mathcal{C}$, $\text{Ind}\,\mathcal{C}$ can be described as the category of all filtered colimits of representable functors in $\text{Set}^{\mathcal{C}^{op}}$, and the functor E_{Ind} is the codomain restriction of the Yoneda embedding.

2. One can proceed analogously with sifted colimits: we denote the free completion under sifted colimits by

$$E_{Sind}: \mathcal{C} \to \text{Sind}\,\mathcal{C}.$$

For a small category $\mathcal{C}$, $\text{Sind}\,\mathcal{C}$ can be described as the category of all sifted colimits of representable functors in $\text{Set}^{\mathcal{C}^{op}}$, and E_{Sind} is the codomain restriction of the Yoneda embedding. We are not going to prove these results in full generality here, the interested reader can find them in Adámek and Rosický (2001). We only prove point 2 under the assumption that $\mathcal{C}$ has finite coproducts, and we sketch the proof of point 1 under the assumption that $\mathcal{C}$ has finite colimits.

4.13 Theorem *For every algebraic theory $\mathcal{T}$, the Yoneda embedding*

$$Y_{\mathcal{T}}: \mathcal{T}^{op} \to \text{Alg}\,\mathcal{T}$$

is a free completion of $\mathcal{T}^{op}$ under sifted colimits. In other words,

$$\text{Alg}\,\mathcal{T} = \text{Sind}\,(\mathcal{T}^{op}).$$

Analogously to Remark 4.11, we have, for every functor $F: \mathcal{T}^{op} \to \mathcal{B}$, where $\mathcal{B}$ has sifted colimits, a choice of a sifted colimit preserving functor $F^*: \text{Alg}\,\mathcal{T} \to \mathcal{B}$ satisfying $F = F^* \cdot Y_{\mathcal{T}}$.

Proof This is analogous to the proof of Theorem 4.10 with $\mathcal{T} = \mathcal{C}^{op}$. Given a functor $F: \mathcal{T}^{op} \to \mathcal{B}$ where $\mathcal{B}$ has sifted colimits, we prove that there exists an essentially unique functor

$$F^*: \text{Alg}\,\mathcal{T} \to \mathcal{B}$$

preserving sifted colimits such that $F = F^* \cdot Y_{\mathcal{T}}$. By 0.14, we are forced to define F^* on objects $A = \text{colim}\,(Y_{\mathcal{T}} \cdot \Phi_A)$ by

$$F^*A = \underset{El\,A}{\text{colim}}\,(F \cdot \Phi_A).$$

This definition makes sense because by Lemma 4.2, $El\,A$ is sifted. As in Theorem 4.10, all that needs to be proved is that the resulting functor F^* preserves sifted colimits. In the present situation, F^* does not have a right adjoint. Nevertheless, since the inclusion $I: \text{Alg}\,\mathcal{T} \to \text{Set}^{\mathcal{T}}$ preserves sifted colimits (2.5), we still have, for $RB = \mathcal{B}(F-, B)$, a bijection

$$\mathcal{B}(F^*A, B) \simeq \text{Set}^{\mathcal{T}}(IA, RB)$$

natural in $A \colon \mathcal{C}^{op} \to Set$ and $B \in \mathcal{B}$. The argument is analogous to Theorem 4.10: both sides represent cocones of the (sifted) diagram $F \cdot \Phi_A$ with codomain B. From the preceding natural bijection, one deduces that F^* preserves sifted colimits: for every fixed object B, the functor $A \mapsto \mathcal{B}(F^*A, B)$ preserves sifted colimits and we now use 0.12. $\square$

4.14 Corollary *A category $\mathcal{A}$ is algebraic iff it is a free completion of a small category with finite coproducts under sifted colimits.*

4.15 Remark Let $\mathcal{T}$ be an algebraic theory. If $\mathcal{B}$ is cocomplete and the functor $F \colon \mathcal{T}^{op} \to \mathcal{B}$ preserves finite coproducts, then its extension

$$F^* \colon Alg\, \mathcal{T} \to \mathcal{B}$$

preserving sifted colimits has a right adjoint. In fact, since F preserves finite coproducts, the functor $B \mapsto \mathcal{B}(F-, B)$ factorizes through $Alg\, \mathcal{T}$, and the resulting functor

$$R \colon \mathcal{B} \to Alg\, \mathcal{T}, \quad B \mapsto \mathcal{B}(F-, B)$$

is a right adjoint to F^*.

4.16 Remark Let $\mathcal{T}$ be a finitely complete small category, and $Lex\, \mathcal{T}$ denote the full subcategory of $Set^{\mathcal{T}}$ of finite limits preserving functors.

1. $Y_{\mathcal{T}} \colon \mathcal{T}^{op} \to Lex\, \mathcal{T}$ preserves finite colimits.
2. The embedding $Lex\, \mathcal{T} \to Set^{\mathcal{T}}$ preserves limits and filtered colimits.
3. $Lex\, \mathcal{T}$ is cocomplete.

The proofs of points 1 and 2 are easy modifications of 1.13, 1.21, and 2.5. Using Remark 4.3, the proof of point 3 is analogous to that of Theorem 4.5.

4.17 Theorem *For every finitely complete small category $\mathcal{T}$, the Yoneda embedding*

$$Y_{\mathcal{T}} \colon \mathcal{T}^{op} \to Lex\, \mathcal{T}$$

is a free completion of $\mathcal{T}^{op}$ under filtered colimits. In other words,

$$Lex\, \mathcal{T} = Ind\,(\mathcal{T}^{op}).$$

Proof A functor $A \colon \mathcal{T} \to Set$ preserves finite limits iff $El\, A$ is filtered (Remark 4.3). Moreover, the embedding $Lex\, \mathcal{T} \to Set^{\mathcal{T}}$ preserves filtered colimits (Remark 4.16). The rest of the proof is a trivial modification of the proof of Theorem 4.13: just replace *sifted* with *filtered* everywhere. $\square$

4.18 Remark Let T be a finitely complete small category. Analogously to Remark 4.15, if B is cocomplete and the functor $F: T^{op} \to B$ preserves finite colimits, then its extension $F^*: \text{Lex}\,T \to B$ preserving filtered colimits has a right adjoint.

Historical remarks

Free colimit completions (see Theorem 4.10) were probably first described by Ulmer (1968); see also Gabriel and Ulmer (1971). The completion *Ind* was introduced by Artin et al. (1972), but it is also contained in Gabriel and Ulmer (1971). The completion *Sind* was introduced in Adámek and Rosický (2001), together with its relation to algebraic categories. But a general completion under a class of colimits is already treated in Gabriel and Ulmer (1971). Later, these completions were studied by a number of authors; see, for example, the results and the references in Adámek et al. (2002) and Kelly (1982).

5
Properties of algebras

In classical algebra, a number of abstract properties of algebras play a central role, for example, *finite presentability*: this means that the algebra can be presented (up to isomorphism) by finitely many generators and equations. However, this is a definite categorical property because it is satisfied by precisely those algebras whose hom-functor preserves filtered colimits (see Chapter 11). Another important concept is that of regular projective algebra (see the next definition). The combination of the preceding two properties is called *perfect presentability*, and we prove that perfectly presentable algebras are precisely those algebras whose hom-functor preserves sifted colimits.

5.1 Definition An object A of a category $\mathcal{A}$ is called *regular projective* if its hom-functor $\mathcal{A}(A, -): \mathcal{A} \to Set$ preserves regular epimorphisms; that is, for every regular epimorphism $e: X \to Z$ and every morphism $f: A \to Z$, there exists a commutative triangle

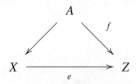

5.2 Example

1. In *Set*, all objects are regular projective.
2. We will see in 11.26 that every free algebra is regular projective.
3. Every regular projective abelian group is free: express A as a regular quotient $e: X \to A$ of a free abelian group X and apply the previous definition to $f = \mathrm{id}_A$. This shows that A is isomorphic to a subgroup of a free abelian group,

and therefore it is free. The same argument holds for regular projective groups and for regular projective Lie algebras.
4. Every finite boolean algebra (not only the free ones) is regular projective, being a retract of a free boolean algebra, see Lemma 5.9.
5. A graph G is regular projective in *Graph* (see 1.11) iff its edges are pairwise disjoint; that is, G is a coproduct of vertices and edges.

5.3 Definition Let $\mathcal{A}$ be a category. An object A of $\mathcal{A}$ is

1. *finitely presentable* if the hom-functor $\mathcal{A}(A, -)\colon \mathcal{A} \to \mathit{Set}$ preserves filtered colimits
2. *perfectly presentable* if the hom-functor $\mathcal{A}(A, -)\colon \mathcal{A} \to \mathit{Set}$ preserves sifted colimits.

5.4 Remark Any perfectly presentable object is finitely presentable (because filtered colimits are sifted) and, assuming the existence of kernel pairs, regular projective (because regular epimorphisms are, by Example 3.4, coequalizers of reflexive pairs). We will show in Corollary 5.16 that in an algebraic category, also the converse implication holds: perfectly presentable objects are precisely the finitely presentable regular projectives. In fact, the converse implication holds in any cocomplete exact category (see 18.3).

5.5 Example Let $\mathcal{T}$ be a small category.

1. In $\mathit{Set}^{\mathcal{T}}$, the representable functors are perfectly presentable, in fact, they have a stronger property: their hom-functors preserve *all* colimits. This follows from the Yoneda lemma and the fact that in $\mathit{Set}^{\mathcal{T}}$, colimits are formed objectwise.
2. If $\mathcal{T}$ is an algebraic theory, representable functors are perfectly presentable objects in $\mathit{Alg}\,\mathcal{T}$. This follows from point 1 and the fact that $\mathit{Alg}\,\mathcal{T}$ is closed in $\mathit{Set}^{\mathcal{T}}$ under sifted colimits (see 2.5).
3. Analogously, if $\mathcal{T}$ has finite limits, then representable functors are finitely presentable objects in $\mathit{Lex}\,\mathcal{T}$. This follows from point 1 and the fact that $\mathit{Lex}\,\mathcal{T}$ is closed in $\mathit{Set}^{\mathcal{T}}$ under filtered colimits (see 4.16).

5.6 Example

1. Every finite set is perfectly presentable in *Set*.
2. An abelian group A is finitely presentable in the preceding sense in the category *Ab* iff it is finitely presentable in the usual algebraic sense; that is, A can be presented by finitely many generators and finitely many equations. This is easily seen from the fact that every abelian group is a filtered colimit of abelian groups that are finitely presentable (in the algebraic

sense). An abelian group is perfectly presentable iff it is free on finitely many generators.
3. In a poset considered as a category, the finitely (or perfectly) presentable objects are precisely the *compact* elements x, that is, such that for every directed join $y = \bigvee_{i \in I} y_i$, from $x \leq y$ it follows that $x \leq y_i$ for some $i \in I$.
4. A graph is finitely presentable in *Graph* iff it has finitely many vertices and finitely many edges. In fact, it is easy to see that for each such a graph G, the hom-functor $Graph(G, -)$ preserves filtered colimits. Conversely, if G is finitely presentable, use the fact that G is a filtered colimit of all its subgraphs on finitely many vertices and finitely many edges. A graph is perfectly presentable iff it has finitely many vertices and finitely many pairwise disjoint edges.

5.7 Remark We will see in Chapter 11 that the situation described for *Ab* in the preceding example is a special case of the general fact that

1. finite presentability has in algebraic categories the usual algebraic meaning (finitely presentable objects are precisely those which can be, in the classical sense, presented by finitely many generators and finitely many equations)
2. every free algebra is regular projective
3. perfectly presentable algebras are just the retracts of the free algebras on finitely many generators

5.8 Remark As pointed out in Example 5.5, in categories $Set^\mathcal{C}$, the representable objects have the property that their hom-functors preserve all colimits. We call such objects *absolutely presentable*. In algebraic categories, absolutely presentable objects are typically rare. For example, no abelian group A is absolutely presentable: for the initial object 1, the object $Ab(A, 1)$ is never initial in *Set*. However, the categories $Set^\mathcal{C}$ are an exception: every object is a colimit of absolutely presentable objects.

5.9 Lemma *If an object is regular projective (or finitely presentable or perfectly presentable), then every retract has that property too.*

Proof If $f: B \to A$ and $g: A \to B$ are such that $g \cdot f = \mathrm{id}_B$, then for $F = \mathcal{A}(A, -)$ and $G = \mathcal{A}(B, -)$ the natural transformations

$$\alpha = \mathcal{A}(g, -): G \to F \quad \text{and} \quad \beta = \mathcal{A}(f, -): F \to G$$

fulfill $\beta \cdot \alpha = G$. Therefore

$$F \underset{F}{\overset{\alpha \cdot \beta}{\rightrightarrows}} F \xrightarrow{\beta} G$$

is a coequalizer. By interchange of colimits, G preserves every colimit preserved by F. □

5.10 Remark Absolutely presentable objects in Set^C are precisely the retracts of the representable functors.

5.11 Lemma

1. *Perfectly presentable objects are closed under finite coproducts.*
2. *Finitely presentable objects are closed under finite colimits.*

Proof Let us prove the first statement (the proof of the second one is similar). Consider a finite family $(A_i)_{i \in I}$ of perfectly presentable objects. Since $\mathcal{A}(\coprod_I A_i, -) \simeq \prod_I \mathcal{A}(A_i, -)$, the claim follows from the obvious fact that a finite product of functors $\mathcal{A} \to Set$ preserving sifted colimits also preserves them. □

5.12 Corollary *Every object of an algebraic category is*

1. *a sifted colimit of perfectly presentable algebras*
2. *a filtered colimit of finitely presentable algebras*

In fact, point 1 follows from 4.2 and 5.5.2, and point 2 follows from 4.2, 2.21, and 5.11.

5.13 Lemma *Regular projective objects are closed under coproducts.*

Proof Let $(A_i)_{i \in I}$ be a family of regular projective objects and let $e \colon X \to Z$ be a regular epimorphism. The claim follows from the formula $\mathcal{A}(\coprod_I A_i, e) \simeq \prod_I \mathcal{A}(A_i, e)$ and the fact that in Set, regular epimorphisms are stable under products (3.21). □

5.14 Corollary *In every category $Alg\,\mathcal{T}$,*

1. *the perfectly presentable objects are precisely the retracts of representable algebras*
2. *the regular projective objects are precisely the retracts of coproducts of representable algebras*

Proof
1. Following Example 5.5 and Lemma 5.9, a retract of a representable algebra is perfectly presentable. Conversely, due to Lemma 4.2, we can express every perfectly presentable algebra A as a sifted colimit of representable algebras. Since $Alg\,\mathcal{T}(A, -)$ preserves this colimit, it follows that id_A factorizes through some of the colimit morphism $e\colon Y_{\mathcal{T}}(t) \to A$. Thus e is a split epimorphism, and A is a retract of $Y_{\mathcal{T}}(t)$.

2. Following Example 5.5, Example 5.9, and Lemma 5.13, a retract of a coproduct of representable algebras is regular projective. Conversely, due to 4.2, we can express every algebra A as a colimit of representable algebras. Since $Alg\,\mathcal{T}$ is cocomplete, this implies, by 0.7, that A is a regular quotient of a coproduct of representable algebras

$$e\colon \coprod_{i \in I} \mathcal{T}(t_i, -) \to A. \tag{5.1}$$

Therefore, if A is regular projective, it is a retract of $\coprod_{i \in I} \mathcal{T}(t_i, -)$. □

5.15 Corollary *Every algebraic category has enough regular projective objects; that is, every algebra is a regular quotient of a regular projective algebra.*

In fact, the morphism e from the preceding proof is a regular epimorphism and, following Example 5.5 and Lemma 5.13, $\coprod_{i \in I} \mathcal{T}(t_i, -)$ is a regular projective object.

5.16 Corollary *In an algebraic category, an algebra is perfectly presentable iff it is finitely presentable and regular projective.*

Proof One implication holds in any category (see Remark 5.4). Conversely, if P is a regular projective object in $Alg\,\mathcal{T}$, due to Corollary 5.14.2, P is a retract of a coproduct of representable algebras. Since every coproduct is a filtered colimit of its finite subcoproducts (2.21), if P is also finitely presentable, then it is a retract of a finite coproduct of representable algebras. Following Example 5.5, Lemma 5.9, and Lemma 5.11.1, P is perfectly presentable. □

5.17 Proposition *In every algebraic category, the finitely presentable algebras are precisely the coequalizers of reflexive pairs of homomorphisms between representable algebras.*

Proof One implication is obvious: representable algebras are finitely presentable (Example 5.5), and finitely presentable objects are closed under finite colimits (Lemma 5.11).

Conversely, let A be a finitely presentable algebra. Following 4.2, A is a (sifted) colimit of representable algebras. Thus A is a filtered colimit of finite colimits of representable algebras (see 2.21). From 1.13 and 0.7, we deduce that every finite colimit of representable algebras is a reflexive coequalizer of a parallel pair between two representable algebras. Since A is finitely presentable, it is a retract of one of the preceding coequalizers:

$$Y_T(t_1) \underset{k}{\overset{h}{\rightrightarrows}} Y_T(t_2) \xrightarrow{e} Q \underset{u}{\overset{s}{\rightleftarrows}} A.$$

Here e is a coequalizer of h and k, and $s \cdot u = \mathrm{id}_A$. Since $Y_T(t_2)$ is regular projective (see Remark 5.4 and Example 5.5) and e is a regular epimorphism, there exists a homomorphism $g: Y_T(t_2) \to Y_T(t_2)$ such that $e \cdot g = u \cdot s \cdot e$. Let us observe that the morphism $s \cdot e$ is a joint coequalizer of h, k, and $g \cdot h$:

$$Y_T(t_1) \underset{g \cdot h}{\overset{h}{\underset{k}{\rightrightarrows}}} Y_T(t_2) \xrightarrow{s \cdot e} A.$$

In fact, $s \cdot e \cdot k = s \cdot e \cdot h = s \cdot u \cdot s \cdot e \cdot h = s \cdot e \cdot g \cdot h$. Assume that a morphism $f: Y_T(t_2) \to X$ coequalizes h, k, and $g \cdot h$. Then there exists a unique homomorphism $v: Q \to X$ with $v \cdot e = f$. Hence $f \cdot h = f \cdot g \cdot h = v \cdot e \cdot g \cdot h = v \cdot u \cdot s \cdot e \cdot h$. Since h is an epimorphism (because the pair h, k is reflexive), we get $f = v \cdot u \cdot s \cdot e$. Now, from 0.7, we conclude that A is a reflexive coequalizer of a pair of homomorphisms between finite coproducts of representable algebras. By 1.13, we get the claim. □

5.18 Remark Beside finite presentability, an important concept in general algebra is finite generation: an algebra A is finitely generated if it has a finite subset not contained in any proper subalgebra. (Or, equivalently, A is a regular quotient of a free algebra on finitely many generators.) This concept also has a categorical formulation. For this we need to introduce the following.

5.19 Definition A *directed union* is a filtered colimit of subobjects; that is, given a filtered diagram $D: \mathcal{D} \to \mathcal{A}$ where D maps every morphism to a monomorphism, colim D is called the *directed union*.

5.20 Remark In Set, directed unions have colimit cocones formed by monomorphisms. Thus the same holds in Set^T. Since $Alg\,T$ is closed under filtered colimits in Set^T, this is also true in $Alg\,T$.

5.21 Definition Let $\mathcal{A}$ be a category. An object A of $\mathcal{A}$ is *finitely generated* if the hom-functor $\mathcal{A}(A, -): \mathcal{A} \to Set$ preserves directed unions.

5.22 Proposition *In every algebraic category, the finitely generated algebras are precisely the regular quotients of representable algebras.*

Proof

1. Let A be finitely generated. Recall from 4.2 that A is the sifted colimit of

$$El\, A \xrightarrow{\Phi_A} \mathcal{T}^{op} \xrightarrow{Y_{\mathcal{T}}} Alg\, \mathcal{T}.$$

Let (X, x) be an object of $El\, A$, and consider the regular factorization of the colimit morphism $\hat{x}$:

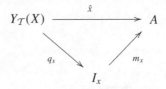

These objects I_x and the connecting monomorphisms between them one gets from the morphisms of $El\, A$ (via diagonal fill-in; see 0.16) form a filtered diagram of monomorphisms. Indeed, given two elements (X, x) and (Z, z) of A, for the element $(X \times Z, (x, z))$, we see that m_x and m_z both factorize through $m_{(x,z)}$. It is clear that $A = colim\, I_x$, and since A is finitely generated, $\mathcal{A}(A, -)$ preserves this colimit. Thus there exists $(X, x) \in El\, A$ and $f: A \to I_x$ such that $id_A = m_x \cdot f$. Therefore m_x, being a monomorphism and a split epimorphism, is an isomorphism. This implies that $\hat{x}: Y_{\mathcal{T}}(X) \to A$ is a regular epimorphism.

2. For every regular epimorphism $e: Y_{\mathcal{T}}(X) \to A$ in $Alg\, \mathcal{T}$, we prove that A is finitely generated. Given a filtered diagram $D: \mathcal{D} \to Alg\, \mathcal{T}$ of subobjects with colimit cocone $c_d: Dd \to C$ ($d \in obj\, \mathcal{D}$), since each c_d is a monomorphism (Remark 5.20), it is sufficient to prove that every morphism $f: A \to C$ factorizes through some c_d. In fact, since $Y_{\mathcal{T}}(X)$ is finitely presentable, the morphism $f \cdot e: Y_{\mathcal{T}}(X) \to C$ factorizes through some c_d – and then we just use the diagonal fill-in:

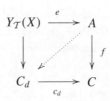

□

5.23 Example In the category $\mathbb{N}/Set$ of sets with countably many constants, the finitely generated objects are those that have, beside the constants, only finitely many elements, whereas the finitely presentable objects have, moreover, the property that only finitely many pairs of distinct natural numbers label the same constant. (Thus, e.g., the terminal object is finitely generated but not finitely presentable.) Finally, the absolutely presentable objects are those finitely generated objects where the constants are pairwise distinct.

Historical remarks

The lecture notes by Gabriel and Ulmer (1971) are the source of the concept of a finitely presentable object. In Adámek and Rosický (2001), perfectly presentable objects were introduced under the name of strongly finitely presentable objects. In algebraic categories, they coincide with objects "projectif-de-type-fini" of Diers (1976) and with the finitely presentable effective projectives of Pedicchio and Wood (2000).

The term *perfectly presentable* was suggested by Joyal (see A. Joyal 2008); his motivation comes from perfect complexes, as explained in 6.11.

6
A characterization of algebraic categories

We have already characterized algebraic categories as free completions (see 4.14). The aim of this chapter is to characterize them as those cocomplete categories that have a strong generator formed by perfectly presentable objects. From that we derive several stability results about algebraic categories: if $\mathcal{A}$ is algebraic, then all slice categories $\mathcal{A} \downarrow A$ and all functor categories $\mathcal{A}^\mathcal{D}$ are algebraic, and so are the reflective subcategories closed under sifted colimits.

6.1 Definition

1. A set of objects $\mathcal{G}$ in a category $\mathcal{A}$ is called a *generator* if two morphisms $x, y \colon A \rightrightarrows B$ are equal whenever $x \cdot g = y \cdot g$ for every morphisms $g \colon G \to A$ with domain G in $\mathcal{G}$.
2. A generator $\mathcal{G}$ is called *strong* if a monomorphism $m \colon A \to B$ is an isomorphism whenever every morphism $g \colon G \to B$ with domain G in $\mathcal{G}$ factorizes through m.

6.2 Remark Here is an equivalent way to express the notions of *generator* and *strong generator*. Consider the functor

$$\mathcal{A} \to Set^\mathcal{G}, \quad A \mapsto \langle \mathcal{A}(G, A) \rangle_{G \in \mathcal{G}}.$$

1. $\mathcal{G}$ is a generator iff the preceding functor is faithful.
2. $\mathcal{G}$ is a strong generator iff the preceding functor is faithful and conservative (see 0.2).

The following proposition suggests that a "strong" generator should more properly be called "extremal"; the present terminology simply has historical roots.

6.3 Proposition *In a category $\mathcal{A}$ with coproducts, a set of objects $\mathcal{G}$ is*

1. *a generator iff every object of $\mathcal{A}$ is a quotient of a coproduct of objects from $\mathcal{G}$*
2. *a strong generator iff every object of $\mathcal{A}$ is an extremal quotient of a coproduct of objects from $\mathcal{G}$.*

Explicitly, $\mathcal{G}$ is a (strong) generator iff for every object A in $\mathcal{A}$, an (extremal) epimorphism

$$e: \coprod_{i \in I} G_i \to A$$

exists with all G_i in $\mathcal{G}$. We will see in the proof that this is equivalent to saying that the *canonical morphism*

$$e_A: \coprod_{(G,g) \in \mathcal{G} \downarrow A} G \to A,$$

whose (G, g) component is g, is an (extremal) epimorphism.

Proof
1. If $\mathcal{G}$ is a generator, then the canonical morphism e_A is obviously an epimorphism. Conversely, if $e: \coprod_{i \in I} G_i \to A$ is an epimorphism, then given distinct morphisms $x, y: A \rightrightarrows B$, some component $e_i: G_i \to A$ fulfils $x \cdot e_i \neq y \cdot e_i$.

2. Let $\mathcal{G}$ be a strong generator. If e_A factorizes through a monomorphism m, then all of its components $g: G \to A$ factorize through m. Since $\mathcal{G}$ is a strong generator, this implies that m is an isomorphism. Conversely, let $e: \coprod_{i \in I} G_i \to A$ be an extremal epimorphism, and consider a monomorphism $m: B \to A$. If every morphism $g: G \to A$ (with G varying in $\mathcal{G}$) factorizes through m, then e factorizes through m so that m is an isomorphism. □

6.4 Corollary *If $\mathcal{A}$ has colimits and every object of $\mathcal{A}$ is a colimit of objects from a set $\mathcal{G}$, then $\mathcal{G}$ is a strong generator.*

In fact, this follows from 0.7 and Proposition 6.3 because any regular epimorphism is extremal.

6.5 Example

1. Every nonempty set forms a (singleton) strong generator in *Set*.
2. The group of integers forms a strong generator in *Ab*.
3. In the category of posets and order-preserving functions, the terminal (one-element) poset forms a generator – but this generator is not strong. In contrast, the two-element chain is a strong generator.

6.6 Example

1. Let $\mathcal{C}$ be a small category. Then $Set^{\mathcal{C}}$ has a strong generator formed by all representable functors. This follows from 0.14 and 6.4.
2. Analogously, for an algebraic theory $\mathcal{T}$, the category $Alg\,\mathcal{T}$ has a strong generator formed by all representable algebras (see 6.4 and 4.2).

6.7 Lemma *Let $\mathcal{A}$ be a cocomplete category with a set of perfectly presentable objects such that every object of $\mathcal{A}$ is a sifted colimit of objects of that set. Then $\mathcal{A}$ has, up to isomorphism, only a set of perfectly presentable objects.*

Proof Express an object A of $\mathcal{A}$ as a sifted colimit of objects from the given set $\mathcal{G}$. If A is perfectly presentable, then it is a retract of an object from $\mathcal{G}$. Clearly each retract of an object B gives rise to an idempotent morphism $e\colon B \to B$, $e \cdot e = e$. Moreover, if two retracts give rise to the same idempotent, then they are isomorphic. Since each object from $\mathcal{A}$ has only a set of retracts, see 0.1, our claim is proved. $\square$

6.8 Remark A result analogous to Lemma 6.7 holds for finitely presentable objects and filtered colimits. The proof is the same.

6.9 Theorem: Characterization of algebraic categories *The following conditions on a category $\mathcal{A}$ are equivalent:*

1. *$\mathcal{A}$ is algebraic.*
2. *$\mathcal{A}$ is cocomplete and has a set $\mathcal{G}$ of perfectly presentable objects such that every object of $\mathcal{A}$ is a sifted colimit of objects of $\mathcal{G}$.*
3. *$\mathcal{A}$ is cocomplete and has a strong generator consisting of perfectly presentable objects.*

Moreover, if the strong generator $\mathcal{G}$ in point 3 is closed under finite coproducts, then the dual of $\mathcal{G}$ (seen as a full subcategory) is an algebraic theory of $\mathcal{A}$.

Proof For the implication $1 \Rightarrow 2$, let $\mathcal{T}$ be an algebraic theory. Then $Alg\,\mathcal{T}$ is cocomplete (4.5), the representable algebras form a set of perfectly presentable objects (5.5), and every algebra is a sifted colimit of representable algebras (4.2).

For the implication $2 \Rightarrow 3$, consider the family $\mathcal{A}_{pp}$ of all perfectly presentable objects of $\mathcal{A}$. By Lemma 6.7, $\mathcal{A}_{pp}$ is essentially a set. By Lemma 4.2 and Corollary 6.4, it is a strong generator.

For the implication $3 \Rightarrow 1$, let $\mathcal{G}$ be a strong generator consisting of perfectly presentable objects. Since perfectly presentable objects are closed under finite coproducts (5.11), we can assume without loss of generality that $\mathcal{G}$ is closed

A characterization of algebraic categories

under finite coproducts (if this is not the case, we can replace $\mathcal{G}$ by its closure in $\mathcal{A}$ under finite coproducts, which still is a strong generator). We are going to prove that $\mathcal{A}$ is equivalent to $Alg(\mathcal{G}^{op})$, where $\mathcal{G}$ is seen as a full subcategory of $\mathcal{A}$.

1. We prove first that $\mathcal{G}$ is *dense*; that is, for every object K of $\mathcal{A}$, the canonical diagram of all morphisms from $\mathcal{G}$,

$$D_K: \mathcal{G} \downarrow K \to \mathcal{A}, \quad (g: G \to K) \mapsto G,$$

has K as colimit, with $(g: G \to K)$ as colimit cocone. To prove this, form a colimit cocone of D_K:

$$(c_g: G \to K^*) \quad \text{for all } g: G \to K \text{ in } \mathcal{G} \downarrow K.$$

We have to prove that the unique factorizing morphism $\lambda: K^* \to K$ with $\lambda \cdot c_g = g$ for all g in $\mathcal{G} \downarrow K$ is an isomorphism. Consider the coproduct

$$\coprod_{(G,g) \in \mathcal{G} \downarrow K} G$$

with coproduct injections $\rho_g: G \to \coprod_{(G,g) \in \mathcal{G} \downarrow K} G$. We have a commutative triangle

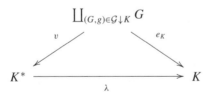

where e_K is the canonical morphism (Proposition 6.3) and v is defined by $v \cdot \rho_g = c_g$ for all $(G, g) \in \mathcal{G} \downarrow K$. Since e_K is an extremal epimorphism (see Proposition 6.3), then λ is an extremal epimorphism. It remains to be proven that λ is a monomorphism. Consider two morphisms $x, x': X \rightrightarrows K^*$ such that $\lambda \cdot x = \lambda \cdot x'$, and let us prove that $x = x'$. While $\mathcal{G}$ is a (strong) generator, we can assume without loss of generality that X is in $\mathcal{G}$. Since $\mathcal{G} \downarrow K$ is sifted (in fact, it has finite coproducts because $\mathcal{G}$ has; see 2.16) and X is perfectly presentable, both x and x' factorize through some colimit morphism; that is, for some (G, g) and (G', g') in $\mathcal{G} \downarrow K$, we have a commutative diagram

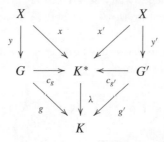

Since, by naturality of c_g, $c_{\lambda \cdot x} = c_g \cdot y = x$ and, analogously, $c_{\lambda \cdot x'} = x'$, we get $x = c_{\lambda \cdot x} = c_{\lambda \cdot x'} = x'$.

2. It follows from point 1 that the functor

$$E: \mathcal{A} \to Alg\,(\mathcal{G}^{op}), \quad K \mapsto \mathcal{A}(-, K)$$

is full and faithful. Indeed, given a homomorphism $\alpha: \mathcal{A}(-, K) \to \mathcal{A}(-, L)$, for every $g: G \to K$ in $\mathcal{G} \downarrow K$ we have a morphism $\alpha_G(g): G \to L$. Those morphisms form a cocone on $\mathcal{G} \downarrow K$ so that there exists a unique morphism $\hat{\alpha}: K \to L$ such that $\hat{\alpha} \cdot g = \alpha_G(g)$ for all g in $\mathcal{G} \downarrow K$. It is easy to check that $E\hat{\alpha} = \alpha$ and that $\widehat{Ef} = f$ for all $f: K \to L$ in $\mathcal{A}$. It remains to prove that E is essentially surjective on objects.

3. Let us prove first that E preserves sifted colimits. Consider a sifted diagram $D: \mathcal{D} \to \mathcal{A}$ with colimit $(h_d: Dd \to H)$. For every object G in $\mathcal{G}$, a colimit of $\mathcal{A}(G, -) \cdot D$ in Set is $\mathcal{A}(G, H)$ with the colimit cocone $\mathcal{A}(G, Dd) \to \mathcal{A}(G, H)$ given by composition with h_d (because $\mathcal{G}$ is formed by perfectly presentable objects). This implies that $(Eh_d: E(Dd) \to EH)$ is a colimit of $E \cdot D$ in $Alg\,(\mathcal{G}^{op})$ (sifted colimits are computed objectwise in $Alg\,(\mathcal{G}^{op})$; see 2.5).

4. It follows from point 3 that E is essentially surjective on objects. In fact, we have the following diagram, commutative up to natural isomorphism:

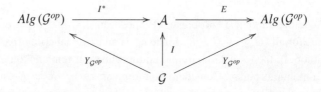

where I is the inclusion and I^* is its extension preserving sifted colimit (4.13). Since $E \cdot I^* \cdot Y_{\mathcal{G}^{op}} \simeq Y_{\mathcal{G}^{op}}$ and $E \cdot I^*$ preserves sifted colimits, it follows from 4.13 that $E \cdot I^*$ is naturally isomorphic to the identity functor. Thus E is essentially surjective on objects. □

A characterization of algebraic categories

6.10 Example

1. The category *Pos* of posets and order-preserving maps is not algebraic: only the discrete posets are perfectly presentable, and there exists no strong generator formed by discrete posets.
2. In the category *Bool* of Boolean algebras, consider the free algebras $\mathcal{PP}n$ on n generators (where $\mathcal{P}X$ is the algebra of all subsets of a set X). The dual of the category formed by all $\mathcal{PP}n$ ($n \in \mathbb{N}$) is an algebraic theory for *Bool*. In fact, $\mathcal{PP}n$ are perfectly presentable and form a strong generator closed under finite coproducts.

6.11 Example Let R be a unitary ring. We denote by $Ch(R)$ the category of chain complexes of left R-modules. Its objects are collections $X = (X_n)_{n \in \mathbb{Z}}$ of left R-modules equipped with a *differential*, that is, a collection of module homomorphisms

$$d = (d_n \colon X_n \to X_{n-1})_{n \in \mathbb{Z}},$$

where $d_{n-1} \cdot d_n = 0$ for each n. Morphisms $f \colon X \to Y$ are *chain maps*, that is, collections $(f_n \colon X_n \to Y_n)_{n \in \mathbb{Z}}$ of module homomorphisms such that $d_n \cdot f_n = f_{n-1} \cdot d_n$ for all n.

A complex X is *bounded* if there are only finitely many $n \in \mathbb{N}$ with $X_n \neq 0$. Since every complex is a filtered colimit of its truncations, each finitely presentable complex is bounded. Perfectly presentable objects in $Ch(R)$ are precisely the bounded complexes of perfectly presentable left R-modules. Since they form a strong generator of $Ch(R)$, the category $Ch(R)$ is algebraic.

6.12 Notation We denote by $\mathcal{A}_{pp}$ a full subcategory of $\mathcal{A}$ representing all perfectly presentable objects of $\mathcal{A}$ up to isomorphism.

6.13 Corollary *For every algebraic category $\mathcal{A}$, the dual of $\mathcal{A}_{pp}$ is an algebraic theory of $\mathcal{A}$: we have an equivalence functor*

$$E \colon \mathcal{A} \to Alg(\mathcal{A}_{pp}^{op}), \quad A \mapsto \mathcal{A}(-, A).$$

In fact, $\mathcal{A}_{pp}$ is a strong generator closed under finite coproducts.

6.14 Corollary *Two algebraic categories $\mathcal{A}$ and $\mathcal{B}$ are equivalent iff the categories $\mathcal{A}_{pp}$ and $\mathcal{B}_{pp}$ are equivalent.*

Proof This follows immediately from Corollary 6.13 and the fact that equivalence functors preserve perfectly presentable objects. □

From 1.5 we know that the slice category $Set \downarrow S$, equivalent to the category Set^S of S-sorted sets, is algebraic. This is a particular case of a more general fact:

6.15 Proposition *Every slice category $\mathcal{A} \downarrow A$ of an algebraic category $\mathcal{A}$ is algebraic.*

Proof The category $\mathcal{A} \downarrow A$ is cocomplete: consider a small category $\mathcal{D}$ and a functor

$$F: \mathcal{D} \to \mathcal{A} \downarrow A, \quad FD = (F_D, f_D: F_D \to A).$$

A colimit of F is given by (C, c), where $(C, \sigma_D: F_D \to C)$ is a colimit of $\Phi_A \cdot F$ (here $\Phi_A: \mathcal{A} \downarrow A \to \mathcal{A}$ is the forgetful functor and $c \cdot \sigma_D = f_D$ for every object D in $\mathcal{D}$). This immediately implies that an object (G, g) is perfectly presentable in $\mathcal{A} \downarrow A$ as soon as G is perfectly presentable in $\mathcal{A}$. Let $\mathcal{G}$ now be a strong generator of $\mathcal{A}$. Then the set of objects $\mathcal{G} \downarrow A = \{(G, g) \mid G \in \mathcal{G}\}$ is a strong generator of $\mathcal{A} \downarrow A$. This is so because a morphism $f: (X, x) \to (Z, z)$ in $\mathcal{A} \downarrow A$ is a strong epimorphism iff $f: X \to Z$ is a strong epimorphism in $\mathcal{A}$. Following Theorem 6.9, $\mathcal{A} \downarrow A$ is algebraic. □

6.16 Lemma *Let the functor $I: \mathcal{A} \to \mathcal{B}$ have a left adjoint R.*

1. *If I is faithful and conservative and $\mathcal{G}$ is a strong generator of $\mathcal{B}$, then $R(\mathcal{G})$ is a strong generator of $\mathcal{A}$.*
2. *If I preserves sifted colimits and X is perfectly presentable in $\mathcal{B}$, then RX is perfectly presentable in $\mathcal{A}$.*

Proof
1. $R(\mathcal{G})$ is a generator because I is a faithful right adjoint. Next, consider a monomorphism $a: A \to A'$ in $\mathcal{A}$ such that every morphism $RG \to A'$ with $G \in \mathcal{G}$ factorizes through a. This implies, by adjunction, that every morphism $G \to IA'$ factorizes through the monomorphism Ia. Since $\mathcal{G}$ is a strong generator, Ia is an isomorphism, and since I is conservative, a is an isomorphism.

2. Since $\mathcal{A}(RX, -) \simeq \mathcal{B}(X, I-) = \mathcal{B}(X, -) \cdot I$, we see that $\mathcal{A}(RX, -)$ is the composite of two functors preserving sifted colimits. □

6.17 Proposition *Let $\mathcal{T}$ be an algebraic theory. Then $Alg\,\mathcal{T}$ is a reflective subcategory of $Set^{\mathcal{T}}$ closed under sifted colimits.*

Proof This is a special case of the adjunction $F^* \dashv R$ for $RB = \mathcal{B}(F-, B)$ obtained in the proof of 4.10. Indeed, by the Yoneda lemma, the right adjoint $Alg\,\mathcal{T}(Y_\mathcal{T}-, -)$ is naturally isomorphic to the full inclusion $Alg\,\mathcal{T} \to Set^{\mathcal{T}}$. □

6.18 Theorem *A category is algebraic iff it is equivalent to a full reflective subcategory of Set^C closed under sifted colimits for some small category C.*

Proof For necessity see 6.17.

Conversely, let C be a small category. By 1.14, Set^C is an algebraic category so that it fulfills the conditions of Theorem 6.9.3. Following Lemma 6.16, those conditions are inherited by any full reflective subcategory closed under sifted colimits of Set^C. $\square$

6.19 Corollary *For any small category D, the functor category A^D of an algebraic category A is algebraic.*

Proof Following Theorem 6.18, there exist a small category C and a full reflection

$$A \xrightarrow[I]{R} Set^C$$

with the right adjoint I preserving sifted colimits. This induces another full reflection

$$A^D \xrightarrow[I\cdot -]{R\cdot -} (Set^C)^D$$

with $I \cdot -$ preserving sifted colimits because by 2.5 they are formed objectwise. Since $(Set^C)^D \simeq Set^{C \times D}$, by Theorem 6.18 A^D is algebraic. $\square$

6.20 Remark Our characterization Theorem 6.9 shows a strong parallel between algebraic categories and the following more general concept of Gabriel and Ulmer (1971).

6.21 Definition A category is called *locally finitely presentable* if it is cocomplete and has a set G of finitely presentable objects such that every object of A is a filtered colimit of objects of G.

6.22 Example

1. Following Theorem 6.9, all algebraic categories are locally finitely presentable.
2. If T is a small category with finite limits, then *Lex* T (see 4.16) is a locally finitely presentable category. In fact, *Lex* T is cocomplete (4.16), the representable functors form a set of finitely presentable objects (5.5), and every object is a filtered colimit of representable functors (4.3).

3. The category *Pos* of posets (which is not algebraic; see Example 6.10) is locally finitely presentable: the two-element chain that forms a strong generator is finitely presentable. Thus we can apply the following.

6.23 Theorem: *Characterization of locally finitely presentable categories. The following conditions on a category $\mathcal{A}$ are equivalent:*

1. *$\mathcal{A}$ is locally finitely presentable.*
2. *$\mathcal{A}$ is equivalent to $\text{Lex}\,\mathcal{T}$ for a small category $\mathcal{T}$ with finite limits.*
3. *$\mathcal{A}$ is cocomplete and has a strong generator formed by finitely presentable objects.*

Moreover, if the strong generator $\mathcal{G}$ in point 3 is closed under finite colimits, then $\mathcal{A}$ is equivalent to $\text{Lex}\,(\mathcal{G}^{op})$.

Proof This proof is quite analogous to that of Theorem 6.9: just change $\text{Alg}\,\mathcal{T}$ to $\text{Lex}\,\mathcal{T}$ for $\mathcal{T} = \mathcal{G}^{op}$; work with finitely presentable objects instead of perfectly presentable ones; and in the proof of $3 \Rightarrow 2$, use the closure under finite colimits. $\square$

6.24 Corollary *A category $\mathcal{A}$ is locally finitely presentable iff it is a free completion of a small, finitely cocomplete category under filtered colimits.*

In fact, this follows from 4.17 and Theorem 6.23.

6.25 Notation We denote by $\mathcal{A}_{fp}$ a full subcategory of $\mathcal{A}$ representing all finitely presentable objects of $\mathcal{A}$ up to isomorphism.

6.26 Corollary *Every locally finitely presentable category $\mathcal{A}$ is equivalent to $\text{Lex}\,\mathcal{T}$ for some small, finitely complete category $\mathcal{T}$. In fact, we have the equivalence functor*

$$E\colon \mathcal{A} \to \text{Lex}\,(\mathcal{A}_{fp}^{op}), \quad A \mapsto \mathcal{A}(-, A).$$

This follows from Theorem 6.23 applied to $\mathcal{G} = \mathcal{A}_{fp}$.

6.27 Proposition *Let $\mathcal{T}$ be a small, finitely complete category. Then $\text{Lex}\,\mathcal{T}$ is a reflective subcategory of $\text{Set}^{\mathcal{T}}$ closed under filtered colimits.*

Proof This is a special case of the adjunction $F^* \dashv R$ for $RB = \mathcal{B}(F-, B)$ obtained in the proof of 4.10. Indeed, by the Yoneda lemma, the right adjoint $\text{Lex}\,\mathcal{T}(Y_{\mathcal{T}}-, -)$ is naturally isomorphic to the full inclusion $\text{Lex}\,\mathcal{T} \to \text{Set}^{\mathcal{T}}$. $\square$

6.28 Theorem *A category is locally finitely presentable iff it is equivalent to a full reflective subcategory of Set^C closed under filtered colimits for some small category C.*

6.29 Corollary *Let T be a small, finitely complete category. Then $Lex\,T$ is a full reflective subcategory of $Alg\,T$ closed under filtered colimits.*

Proof Consider the full inclusions

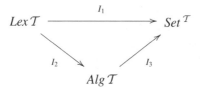

By Proposition 6.27, I_1 has a left adjoint, say, R. Since I_3 is full and faithful, $R \cdot I_3$ is left adjoint to I_2. Finally, I_2 preserves filtered colimits because I_1 preserves them by Proposition 6.27 and I_3 reflects them. □

In 12.12, we will need the following fact.

6.30 Corollary *Let $\mathcal{A}$ be an algebraic category and $\mathcal{B}$ a category with sifted colimits. If in $\mathcal{A}$ every finitely presentable object is regular projective, then a functor $F\colon \mathcal{A} \to \mathcal{B}$ preserving filtered colimits preserves also sifted colimits.*

Proof Following Corollaries 6.13 and 6.26,

$$Alg\,(\mathcal{A}_{pp}^{op}) \simeq \mathcal{A} \simeq Lex\,(\mathcal{A}_{fp}^{op}).$$

If, moreover, finitely presentable objects in $\mathcal{A}$ are regular projective, then $\mathcal{A}_{pp} = \mathcal{A}_{fp}$ (5.16). The result now follows from the universal properties stated in 4.13 and 4.17. □

6.31 Example *Set* (and more generally Set^S) and the category of vector spaces over a field are examples of algebraic categories where every (finitely presentable) object is regular projective. More generally, in the category of left modules over a semisimple ring, every object is regular projective.

Let us finish this chapter by quoting from Bunge (1966) another characterization theorem similar to Theorem 6.9. Recall that Set^C is the free completion of C^{op} under colimits (4.10); recall also the concept of the absolutely presentable object of 5.8.

6.32 Theorem *A category is equivalent to a functor category Set^C for some small category C iff it is cocomplete and has a strong generator consisting of absolutely presentable objects.*

Historical remarks

The concept of the locally finitely presentable category, due to Gabriel and Ulmer (1971), was an attempt toward a categorical approach to categories of finitary structures generalizing Lawvere's algebraic theories. This is the source of Theorem 6.23. See Adámek and Rosický (1994) and Makkai and Paré (1989) for more recent monographs on locally presentable categories and their generalizations. In Adámek and Rosický (2001), the analogy between locally finitely presentable categories and algebraic categories was made explicit.

A characterization of categories of algebras for one-sorted algebraic theories is contained in the thesis of Lawvere (1963). The characterization Theorem 6.9 is taken from Adámek and Rosický (2001), but the equivalence of points 1 and 3 was already proved by Diers (1976). A first characterization of categories of *Set*-valued functors is in the thesis of Bunge (1966) (the first proof was published by Linton, 1969b). There is a general result covering Theorems 6.9, 6.23, and 6.32, see Centazzo et al. (2004).

7
From filtered to sifted

The aim of this chapter is to demonstrate that the "equation"

$$\text{sifted colimits} = \text{filtered colimits} + \text{reflexive coequalizers}$$

is almost valid – but not quite. What we have in mind are three facts:

1. A category $\mathcal{C}$ has sifted colimits iff it has filtered colimits and reflexive coequalizers. This holds whenever $\mathcal{C}$ has finite coproducts – and in general, it is false.
2. A functor preserves sifted colimits iff it preserves filtered colimits and reflexive coequalizers. This holds whenever the domain category is finitely cocomplete – and in general, it is false.
3. The free completion $Sind\,\mathcal{C}$ of a small category $\mathcal{C}$ under sifted colimits is obtained from the free completion $Rec\,\mathcal{C}$ under reflexive coequalizers by completing it under filtered colimits:

$$Sind\,\mathcal{C} = Ind\,(Rec\,\mathcal{C}).$$

This holds whenever $\mathcal{C}$ has finite coproducts – and in general, it is false.

We begin by describing $Rec\,\mathcal{C}$ in a manner analogous to the description of $Sind\,\mathcal{C}$ and $Ind\,\mathcal{C}$ (see 4.13 and 4.17). A quite different approach to $Rec\,\mathcal{C}$ is treated in Chapter 17.

As a special case of Definition 4.9, we get the following.

7.1 Definition By a *free completion of a category $\mathcal{C}$ under reflexive coequalizers* is meant a functor $E_{Rec}\colon \mathcal{C} \to Rec\,\mathcal{C}$ such that

1. $Rec\,\mathcal{C}$ is a category with coequalizers of reflexive pairs

2. for every functor $F\colon \mathcal{C} \to \mathcal{B}$, where $\mathcal{B}$ is a category with reflexive coequalizers, there exists an essentially unique functor $F^*\colon Rec\,\mathcal{C} \to \mathcal{B}$ preserving reflexive coequalizers with F naturally isomorphic to $F^* \cdot E_{Rec}$.

Recall that for a category $\mathcal{A}$, we denote by $\mathcal{A}_{fp}$ the full subcategory of finitely presentable objects.

7.2 Lemma *Let $\mathcal{T}$ be an algebraic theory. The inclusion $I\colon (Alg\,\mathcal{T})_{fp} \to Set^{\mathcal{T}}$ preserves reflexive coequalizers.*

Proof This follows from the fact that $(Alg\,\mathcal{T})_{fp}$ is closed in $Alg\,\mathcal{T}$ under finite colimits (5.11) and that $Alg\,\mathcal{T}$ is closed in $Set^{\mathcal{T}}$ under sifted colimits (2.5); see also 3.2. □

7.3 Theorem *For every algebraic theory $\mathcal{T}$, the restricted Yoneda embedding*

$$Y_{\mathcal{T}}\colon \mathcal{T}^{op} \to (Alg\,\mathcal{T})_{fp}$$

is a free completion of $\mathcal{T}^{op}$ under reflexive coequalizers. In other words,

$$(Alg\,\mathcal{T})_{fp} = Rec\,(\mathcal{T}^{op}).$$

Proof Recall from 3.2 the category $\mathcal{M}$

$$P \xleftarrow{\;\;d\;\;} \begin{array}{c} f_1 \\ \rightrightarrows \\ f_2 \end{array} Q \quad \text{modulo} \quad f_1 \cdot d = \mathrm{id}_Q = f_2 \cdot d.$$

We will prove that given a finitely presentable algebra $A\colon \mathcal{T} \to Set$, there exists a final functor (see Definition 2.12)

$$M\colon \mathcal{M} \to El\,A.$$

The rest of the proof is analogous to the proof of Theorem 4.10: given a functor $F\colon \mathcal{T}^{op} \to \mathcal{B}$ where $\mathcal{B}$ has reflexive coequalizers, we prove that there exists an essentially unique functor

$$F^*\colon (Alg\,\mathcal{T})_{fp} \to \mathcal{B}$$

preserving reflexive coequalizers and such that $F \simeq F^* \cdot Y_{\mathcal{T}}$. In fact, using the notation of 0.14 for the final functor M above, we see that the reflexive coequalizer of $F \cdot \Phi_A \cdot M(f_i)$ ($i = 1, 2$) in $\mathcal{B}$ is just the colimit of $F \cdot \Phi_A$; thus the latter colimit exists, and we are forced to define

$$F^*A = \underset{El\,A}{colim}(F \cdot \Phi_A)$$

on objects. This extends uniquely to morphisms (as in the proof of 4.10) and yields a functor $F^*: (Alg\,\mathcal{T})_{fp} \to \mathcal{B}$. Since the inclusion $I: (Alg\,\mathcal{T})_{fp} \to Set^{\mathcal{T}}$ preserves reflexive coequalizers (Lemma 7.2), we have for $RB = \mathcal{B}(F-, B)$ a bijection

$$\mathcal{B}(F^*A, B) \simeq Set^{\mathcal{T}}(IA, RB)$$

natural in A and B, from which one deduces that F^* preserves reflexive coequalizers.

To prove the existence of the final functor M, recall from 5.17 that there exists a reflexive pair

$$\overline{P} \underset{\overline{f_2}}{\overset{\overline{f_1}}{\rightrightarrows}} \overline{Q}$$
(with $\overline{d}$ in between)

in $\mathcal{T}$ such that A is a coequalizer of $Y_{\mathcal{T}}(\overline{f_i})$:

$$Y_{\mathcal{T}}(\overline{P}) \underset{Y_{\mathcal{T}}(\overline{f_2})}{\overset{Y_{\mathcal{T}}(\overline{f_1})}{\rightrightarrows}} Y_{\mathcal{T}}(\overline{Q}) \xrightarrow{c} A.$$

Put $\overline{c} = c \cdot Y_{\mathcal{T}}(\overline{f_i})$ and define objects of $El\,A$ as follows:

$$MP = (\overline{P}, \overline{c}_{\overline{P}}(\mathrm{id}_{\overline{P}})) \quad MQ = (\overline{Q}, c_{\overline{Q}}(\mathrm{id}_{\overline{Q}})).$$

Since clearly $\overline{f_i}: MP \to MQ$ and $\overline{d}: MQ \to MP$ are morphisms of $El\,A$, we obtain a functor $M = \overline{(-)}: \mathcal{M} \to El\,A$. Let us prove its finality, applying 2.13.3:

1. Every object (X, x) of $El\,A$ has a morphism into MQ. In fact,

$$c_X: \mathcal{T}(\overline{Q}, X) \to AX$$

is an epimorphism; thus for $x \in AX$ there exists $f: \overline{Q} \to X$ with $c_X(f) = x$, which implies that $f: (X, x) \to MQ$ is a morphism of $El\,A$.

2. Given two morphisms of $El\,A$ from (X, x) to MP or MQ, they are connected by a zigzag in the slice category $(X, x) \downarrow M$. In fact, we can restrict ourselves to the codomain MQ: the general case is then solved by composing morphisms with codomain MP by Mf_1.

   Given morphisms

$$h, k: (X, x) \to MQ,$$

we have
$$A(h) \cdot c_Q(\mathrm{id}_Q) = x = A(k) \cdot c_Q(\mathrm{id}_Q),$$
which, because of the naturality of c, yields
$$c_X(h) = c_X(k).$$
Now use the description of coequalizers in *Set* (see 0.6): since c_X is the coequalizer of $Y_{\mathcal{T}}(\overline{f_1})$ and $Y_{\mathcal{T}}(\overline{f_2})$, there is a zigzag of this pair connecting h and k. For example, a zigzag of length 2 is:

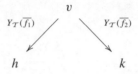

for some $v\colon P \to X$. This means $h = v \cdot f_1$ and $k = v \cdot f_2$ and yields the following zigzag in $(X, x) \downarrow M$:

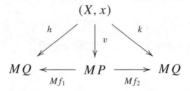

and analogously for longer zigzags. □

7.4 Corollary *For a small category $\mathcal{C}$ with finite coproducts, we have*
$$\mathrm{Sind}\,\mathcal{C} = \mathrm{Ind}\,(\mathrm{Rec}\,\mathcal{C}).$$
More precisely, the composition
$$\mathcal{C} \xrightarrow{E_{Rec}} \mathrm{Rec}\,\mathcal{C} \xrightarrow{E_{Ind}} \mathrm{Ind}\,(\mathrm{Rec}\,\mathcal{C})$$
is a free completion of $\mathcal{C}$ under sifted colimits.

In fact, for $\mathcal{T} = \mathcal{C}^{op}$, the preceding theorem yields $\mathrm{Rec}\,\mathcal{C} = (\mathrm{Alg}\,\mathcal{T})_{fp}$, from which 6.26 and 4.17 prove $\mathrm{Ind}\,(\mathrm{Rec}\,\mathcal{C}) = \mathrm{Alg}\,\mathcal{T}$. Now apply 4.13.

7.5 Remark In the proof of Theorem 7.3, if $\mathcal{B}$ has finite colimits and F preserves finite coproducts, then the extension F^* preserves finite colimits. This follows from the fact that $\mathcal{B}(F-, B)$ lies now in $\mathrm{Alg}\,\mathcal{T}$ and that $(\mathrm{Alg}\,\mathcal{T})_{fp}$ is closed in $\mathrm{Alg}\,\mathcal{T}$ under finite colimits (5.11) so that we have a bijection
$$\mathcal{B}(F^*A, B) \simeq \mathrm{Alg}\,\mathcal{T}(A, \mathcal{B}(F-, B))$$
natural in $A \in (\mathrm{Alg}\,\mathcal{T})_{fp}$ and $B \in \mathcal{B}$.

7.6 Remark In the introduction to this chapter, we claimed that a functor defined on a finitely cocomplete category preserves sifted colimits iff it preserves filtered colimits and reflexive coequalizers. The proof of this result can be found in Adámek et al. (2010); here we present a (simpler) proof based on Corollary 7.4 that requires cocompleteness of both categories (in fact, sifted colimits are enough as far as the codomain category is concerned).

7.7 Theorem *A functor between cocomplete categories preserves sifted colimits iff it preserves filtered colimits and reflexive coequalizers.*

Proof Necessity is clear. To prove sufficiency, let $F: \mathcal{E} \to \mathcal{A}$ preserve filtered colimits and reflexive coequalizers, where $\mathcal{E}$ and $\mathcal{A}$ are cocomplete. For every sifted diagram $D: \mathcal{D} \to \mathcal{E}$, choose a small full subcategory $U: \mathcal{C} \hookrightarrow \mathcal{E}$ containing the image of D and closed in $\mathcal{E}$ under finite coproducts. Consider the following diagram:

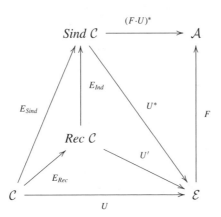

where U^* and $(F \cdot U)^*$ are the extensions of U and $F \cdot U$, respectively, preserving sifted colimits, and U' is the extension of U preserving reflexive coequalizers. Since by Corollary 7.4 we have $E_{Sind} = E_{Ind} \cdot E_{Rec}$, it follows that

$$(F \cdot U)^* \cdot E_{Ind} \cdot E_{Rec} \simeq (F \cdot U)^* \cdot E_{Sind} \simeq F \cdot U \simeq F \cdot U' \cdot E_{Rec}.$$

The functor E_{Ind} preserves reflexive coequalizers by 4.16 and 4.17, and so do the functors F, U' and $(F \cdot U)^*$. Thus the universal property of E_{Rec} yields

$$(F \cdot U)^* \cdot E_{Ind} \simeq F \cdot U'.$$

Since

$$U^* \cdot E_{Ind} \cdot E_{Rec} \simeq U^* \cdot E_{Sind} \simeq U \simeq U' \cdot E_{Rec}$$

and, once again, the functors U^*, E_{Ind}, and U' preserve reflexive coequalizers, we have

$$U^* \cdot E_{Ind} \simeq U'.$$

Finally, since

$$F \cdot U^* \cdot E_{Ind} \simeq F \cdot U' \simeq (F \cdot U)^* \cdot E_{Ind}$$

and the functors F, U^* and $(F \cdot U)^*$ preserve filtered colimits, we have

$$F \cdot U^* \simeq (F \cdot U)^*. \tag{7.1}$$

We are ready to prove $F(colim\, D) \simeq colim\,(F \cdot D)$. Let

$$D' \colon \mathcal{D} \to \mathcal{C} \quad \text{with } D = U \cdot D'$$

be the codomain restriction of D. Then

$$F \cdot D = F \cdot U \cdot D' \simeq (F \cdot U)^* \cdot E_{Sind} \cdot D'$$

implies, since $(F \cdot U)^*$ preserves the sifted colimit of $E_{Sind} \cdot D'$, that

$$colim\,(F \cdot D) \simeq (F \cdot U^*)(colim\,(E_{Sind} \cdot D')).$$

From (7.1) above and the fact that U^* also preserves the sifted colimit of $E_{Sind} \cdot D'$, we derive

$$colim\,(F \cdot D) \simeq F(colim\,(U^* \cdot E_{Sind} \cdot D')) \simeq F(colim\, D).$$

$\square$

7.8 Remark We thus established proofs of the affirmative statements 2 and 3 of the introduction of this chapter. The statement 1 is easy: if $\mathcal{C}$ has filtered colimits, reflexive coequalizers, and finite coproducts, then it has all colimits (4.1). With the following examples, we demonstrate the negative parts of statements 1–3.

7.9 Example Here we give an example of a category not having sifted colimits, although it has (1) filtered colimits and (2) reflexive coequalizers. We start with the following category $\mathcal{D}$ given by the gluing of two reflexive pairs at their codomains; that is, $\mathcal{D}$ is given by the graph

$$A \underset{a_2}{\overset{a_1}{\rightrightarrows}} \overset{d}{\longleftarrow} B \overset{d'}{\longrightarrow} \underset{a'_2}{\overset{a'_1}{\rightleftarrows}} A'$$

and the equations making both parallel pairs reflexive:

$$a_i \cdot d = \mathrm{id}_B = a'_i \cdot d' \quad \text{for } i = 1, 2.$$

The proof that $\mathcal{D}$ is sifted is completely analogous to the proof of 3.2: we verify that colimits over $\mathcal{D}$ in *Set* commute with finite products. Assume that the preceding graph depicts sets A, B and A' and functions between them. Then a colimit can be described as the canonical function $c \colon B \to C = B/\sim$, where two elements $x, y \in B$ are equivalent iff they are connected by a zigzag formed by a_1, a_2, a_1', and a_2'. Since the two pairs are reflexive, the length of the zigzag can be arbitrarily prolonged, and the type can be chosen to be

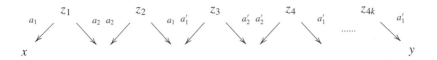

(here, for the elements z_{4i+1}, we use a_1, a_2; for z_{4i+2}, we use a_2, a_1; for z_{4i+3}, we use a_1', a_2'; and for z_{4i}, we use a_2', a_1'). From that it is easy to derive that $\mathcal{D}$ is sifted.

We now add to $\mathcal{D}$ the coequalizers c_1 of a_1, a_2 and c' of a_1', a_2': let $\mathcal{E}$ be the category given by the graph

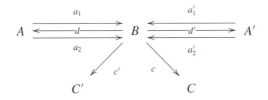

with the previous equations plus the following ones:

$$c \cdot a_1 = c \cdot a_2 \quad \text{and} \quad c' \cdot a_1' = c' \cdot a_2'.$$

The sifted diagram $\mathcal{D} \to \mathcal{E}$, which is the inclusion, does not have a colimit. However, $\mathcal{E}$ has reflexive coequalizers because its only nontrivial reflexive pairs are a_1, a_2 (with coequalizer c) and a_1', a_2' (with coequalizer c'). Moreover, $\mathcal{E}$ has filtered colimits: since the category $\mathcal{E}$ is clearly finite, it does not have any nontrivial filtered diagram, except those obtained by iterating an idempotent endomorphism e (see 2.3). Thus it is sufficient to verify that $\mathcal{E}$ has coequalizers of all pairs e, id_X, where e is an idempotent endomorhism of X. In fact, the only idempotents of $\mathcal{E}$ are $d \cdot a_i$ and $d' \cdot a_i'$. The coequalizer of $d \cdot a_1$ and id_A is clearly a_1 (because a morphism f with $f = f \cdot d \cdot a_1$ fulfills $f \ne a_2$ and then uniquely factorizes through a_1), and analogously for the other three idempotents. Thus the preceding coequalizers demonstrate that $\mathcal{E}$ has filtered colimits.

7.10 Remark For the category $\mathcal{D}$ of Example 7.9, we have

$$Sind\,\mathcal{D} \neq Ind\,(Rec\,\mathcal{D}).$$

Observe first that

$$Rec\,\mathcal{D} = \mathcal{E}$$

is obtained from $\mathcal{D}$ by freely adding the coequalizers c and c'. The category $Ind\,\mathcal{E}$ does not have a terminal object. In fact, the full subcategory of all objects X in $Ind\,\mathcal{E}$ having a morphism from at most one of the objects C or C' into X is clearly closed under filtered colimits – thus every object X has that property. In contrast, $Sind\,\mathcal{D}$ has the terminal object $colim\,\mathcal{D}$.

7.11 Example Here we give an example of a functor not preserving sifted colimits, although it preserves (1) filtered colimits and (2) reflexive coequalizers. Let $\mathcal{E}^T$ be the category $\mathcal{E}$ above with a terminal object T added. The sifted diagram $D\colon \mathcal{D} \to \mathcal{E}^T$, which is the inclusion, has colimit

$$T = colim\,D$$

in $\mathcal{E}^T$. Let $\mathcal{A}$ be the category $\mathcal{E}^T$ with a new terminal object S added. The functor

$$F\colon \mathcal{E}^T \to \mathcal{A} \quad \text{with} \quad FT = S \quad \text{and} \quad FX = X \quad \text{for all } X \neq T,$$

which is the identity map on morphisms of $\mathcal{E}$, does not preserve sifted colimits because

$$F(colim\,D) = S \quad \text{and} \quad colim\,F \cdot D = T.$$

However, F clearly preserves filtered colimits and reflexive coequalizers: the only nontrivial colimits of these types in $\mathcal{E}^T$ lie in $\mathcal{E}$ and are described in Example 7.9. The same description applies to $\mathcal{A}$.

7.12 Example Since $\mathcal{N}$ is a theory of *Set* (see 1.4), we have

$$Set = Ind\,Rec\,\mathcal{N}.$$

Let us observe that in the opposite direction, we do not obtain an equality: the canonical morphism from $Rec\,Ind\,\mathcal{N}$ to *Set* is not an equivalence. In fact, it is easy to see that $Ind\,\mathcal{N}$ can be represented as *Set*. And the free completion $Rec\,Set$ of *Set* under reflexive coequalizers is not an equivalence. To demonstrate this, it is sufficient to present an arbitrary functor with domain *Set* that does not preserve reflexive coequalizers.

One such example is $Set(\mathbb{N}, -)$. In fact, consider the parallel pair

$$\mathbb{N} + \mathbb{N} \xrightarrow[v]{u} \mathbb{N}$$

having the left-hand components $\mathrm{id}_{\mathbb{N}}$ and the right-hand ones $\mathrm{id}_{\mathbb{N}}$ and s, the successor function, respectively. Because of the left-hand components, this is a reflexive pair; because of the right-hand components, their coequalizer has then terminal codomain 1. However, the coequalizer of $u^{\mathbb{N}}$, $v^{\mathbb{N}}$ does not merge the elements $(0, 0, 0, \ldots)$ and $(0, 1, 2, \ldots)$ of $\mathbb{N}^{\mathbb{N}}$; therefore its codomain is not $1^{\mathbb{N}} \simeq 1$.

Historical remarks

The open problem of Adámek and Rosický (2001), whether preservation of filtered colimits and reflexive coequalizers implies preservation of sifted colimits, was answered by Joyal; his proof even works for quasicategories (see Joyal 2008). Another proof was given by Lack (see Lack and Rosický, 2010). The present proof is taken from Adámek et al. (2010), where also the stronger statements stated at the beginning of this chapter are proved. The formula in Corollary 7.4 stems from Adámek and Rosický (2001).

8
Canonical theories

Every algebraic category has a number of algebraic theories that are often nonequivalent; we will study this in more detail in Chapter 15. In the present chapter, we prove that there is always an essentially unique algebraic theory with split idempotents. We call it a canonical theory. This is the idempotent completion (also known as Cauchy completion) of any algebraic theory of the given category. We first discuss splitting of idempotents and idempotent completions.

8.1 Definition

1. Given an idempotent morphism

$$f: X \to X, \quad f \cdot f = f$$

   in a category $\mathcal{C}$, by a *splitting* of f is meant a factorization $f = m \cdot e$ such that $e \cdot m$ is the identity morphism:

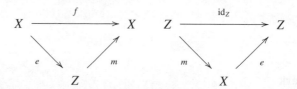

2. A category $\mathcal{C}$ is called *idempotent complete* provided that every idempotent in $\mathcal{C}$ has a splitting.

8.2 Remark

1. A splitting of an idempotent f is unique up to isomorphism:
   a. For every isomorphism $i: Z \to \bar{Z}$, the morphisms $\bar{e} = i \cdot e$ and $\bar{m} = m \cdot i^{-1}$ form a splitting of f.

b. For every splitting $f = \bar{m} \cdot \bar{e}$, $\bar{e} \cdot \bar{m} = \mathrm{id}$, there exists a unique isomorphism i such that $i \cdot e = \bar{e}$ and $m \cdot i^{-1} = \bar{m}$ (just put $i = \bar{e} \cdot m$ and $i^{-1} = e \cdot \bar{m}$).
2. To be idempotent complete is a self-dual notion: $\mathcal{C}$ is idempotent complete iff $\mathcal{C}^{op}$ is so.

8.3 Example

1. Every category that has equalizers is idempotent complete. In fact, consider an equalizer m of the idempotent $f\colon X \to X$ and $\mathrm{id}\colon X \to X$:

$$Z \xrightarrow{m} X \xrightarrow[\mathrm{id}]{f} X.$$

Since $f \cdot f = \mathrm{id} \cdot f$, the morphism f factorizes as $f = m \cdot e$ for some $e\colon X \to Z$. Now $m \cdot e \cdot m = f \cdot m = m$, and m is a monomorphism so that $e \cdot m = \mathrm{id}$. Conversely, if an idempotent $f\colon X \to X$ splits as $f = m \cdot e$, then m is an equalizer of f and id_X.
2. Every category with coequalizers is also idempotent complete. Conversely, if an idempotent $f\colon X \to X$ splits as $f = m \cdot e$, then e is a coequalizer of f and id_X. This is the dualization of 1.
3. A full subcategory of an idempotent-complete category $\mathcal{C}$ is idempotent-complete iff it is closed in $\mathcal{C}$ under retracts.

8.4 Definition
By an *idempotent completion of a category* $\mathcal{C}$ is meant a functor

$$E_{Ic}\colon \mathcal{C} \to Ic\,\mathcal{C}$$

such that

1. $Ic\,\mathcal{C}$ is idempotent complete
2. for every functor $F\colon \mathcal{C} \to \mathcal{B}$, where $\mathcal{B}$ is an idempotent-complete category, there exists an essentially unique functor $F^*\colon Ic\,\mathcal{C} \to \mathcal{B}$, with F naturally isomorphic to $F^* \cdot E_{Ic}$.

8.5 Remark

1. A category $\mathcal{C}$ is idempotent complete iff the functor $E_{Ic}\colon \mathcal{C} \to Ic\,\mathcal{C}$ is an equivalence.
2. Clearly $Ic\,(\mathcal{C}^{op}) \simeq (Ic\,\mathcal{C})^{op}$.

We give now an elementary description of $Ic\,\mathcal{C}$.

8.6 Definition
For every category $\mathcal{C}$, we denote by $Ic\,\mathcal{C}$ the category of idempotents: its objects are the idempotent morphisms of $\mathcal{C}$. Its morphisms

from $f: X \to X$ to $g: Z \to Z$ are the morphisms $a: X \to Z$ in $\mathcal{C}$ such that $a = g \cdot a \cdot f$ or, equivalently, such that the diagram

commutes. The identity of the object $f: X \to X$ is f itself, and composition is as in $\mathcal{C}$. We have a full and faithful functor $E_{Ic}: \mathcal{C} \to Ic\,\mathcal{C}$ defined by $E_{Ic}(X) = \mathrm{id}_X$ and $E_{Ic}(a) = a$.

8.7 Proposition *The functor $E_{Ic}: \mathcal{C} \to Ic\,\mathcal{C}$ defined in 8.6 is an idempotent completion of $\mathcal{C}$.*

Proof

1. $Ic\,\mathcal{C}$ is idempotent complete: let a be an idempotent endomorphism of $(f: X \to X)$ in $Ic\,\mathcal{C}$. A splitting of a is given by

$$(f: X \to X) \xrightarrow{a} (a: X \to X) \xrightarrow{a} (f: X \to X).$$

2. Consider a functor $F: \mathcal{C} \to \mathcal{B}$ with $\mathcal{B}$ idempotent complete. Every object $(f: X \to X)$ of $Ic\,\mathcal{C}$ is obtained by splitting the idempotent $f: E_{Ic}(X) \to E_{Ic}(X)$. Thus we are forced to define F^* on objects $f: X \to X$ as the (essentially unique) splitting of $Ff: FX \to FX$ in $\mathcal{B}$.

To define F^* on morphisms $a: (f: X \to X) \to (g: Z \to Z)$ in $Ic\,\mathcal{C}$, observe that a is the unique morphism making the diagram

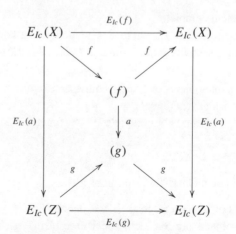

commutative in $Ic\,\mathcal{C}$. Thus we are forced to define $F^*(a)$ as the unique morphism making the diagram

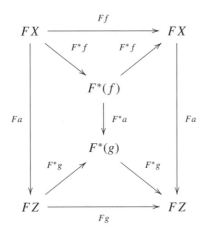

commutative in $\mathcal{B}$. Explicitly;

$$F^*a: \quad F^*(f) \xrightarrow{F^*f} FX \xrightarrow{Fa} FZ \xrightarrow{F^*g} F^*(g).$$

It is easy to verify that this yields a well-defined functor F^* with $F^* \cdot I \simeq F$. $\square$

8.8 Proposition *For every small category $\mathcal{C}$, an idempotent completion of $\mathcal{C}^{op}$ is the codomain restriction of the Yoneda embedding $Y_\mathcal{C}: \mathcal{C}^{op} \to Set^\mathcal{C}$ to the full subcategory of all absolutely presentable objects of $Set^\mathcal{C}$.*

Proof Recall from 5.10 that a functor $\mathcal{C} \to Set$ is absolutely presentable iff it is a retract of a representable functor. To prove that the full subcategory of all absolutely presentable objects of $Set^\mathcal{C}$ is equivalent to the category $Ic\,(\mathcal{C}^{op})$ of Definition 8.6, consider two retracts,

$$R \xrightleftharpoons[m]{e} Y_\mathcal{C}(X), \quad e \cdot m = \mathrm{id}_R \quad \text{and} \quad S \xrightleftharpoons[n]{f} Y_\mathcal{C}(Z), \quad f \cdot n = \mathrm{id}_S,$$

and a morphism $g: R \to S$ in $Set^\mathcal{C}$. By the Yoneda lemma, we get two idempotents $\hat{e} = m \cdot e(\mathrm{id}_X)$ and $\hat{f} = n \cdot f(\mathrm{id}_Z)$ in $\mathcal{C}$ and a morphism $\hat{g} = n \cdot g \cdot e(\mathrm{id}_X)$

forming a commutative diagram

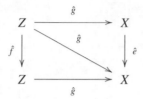

This is a morphism in the category $Ic(\mathcal{C}^{op})$. The rest of the proof is straightforward. □

8.9 Corollary *For every algebraic theory* $\mathcal{T}$, *the Yoneda embedding*

$$Y_{\mathcal{T}}: \mathcal{T}^{op} \to (Alg\,\mathcal{T})_{pp}$$

is an idempotent completion of $\mathcal{T}^{op}$. *In other words,* $(Alg\,\mathcal{T})_{pp} \simeq Ic(\mathcal{T}^{op})$.

In fact, by 5.10 and Proposition 8.8, $Ic(\mathcal{T}^{op})$ is formed by retracts of representables in $Set^{\mathcal{T}}$, that is, by perfectly presentable objects in $Alg\,\mathcal{T}$ (see 5.14).

8.10 Corollary *For two small categories* $\mathcal{C}$ *and* $\mathcal{D}$, *the corresponding functor categories* $Set^{\mathcal{C}}$ *and* $Set^{\mathcal{D}}$ *are equivalent iff* $\mathcal{C}$ *and* $\mathcal{D}$ *have a common idempotent completion:*

$$Set^{\mathcal{C}} \simeq Set^{\mathcal{D}} \quad \text{iff} \quad Ic\,\mathcal{C} \simeq Ic\,\mathcal{D}.$$

Proof The universal property of $E_{Ic}: \mathcal{C} \to Ic\,\mathcal{C}$ clearly implies $Set^{\mathcal{C}} \simeq Set^{Ic\,\mathcal{C}}$. Conversely, if $Set^{\mathcal{C}} \simeq Set^{\mathcal{D}}$, then the subcategories of absolutely presentable objects are equivalent. By 5.10 and 8.8, this means that $Ic(\mathcal{C}^{op}) \simeq Ic(\mathcal{D}^{op})$. Now use duality, or Remark 8.5.2. □

8.11 Definition Let $\mathcal{A}$ be an algebraic category. An algebraic theory for $\mathcal{A}$ is called *canonical* if it is idempotent complete.

8.12 Proposition *Every algebraic category* $\mathcal{A}$ *has a canonical theory unique up to equivalence. The dual of* $\mathcal{A}_{pp}$ *(see 6.12) is a canonical theory for* $\mathcal{A}$.

Proof Following 6.13, $\mathcal{A} \simeq Alg(\mathcal{A}_{pp}^{op})$. By 5.14 and 8.3, $\mathcal{A}_{pp}$ is idempotent-complete, and then $\mathcal{A}_{pp}^{op}$ is also idempotent complete (see Remark 8.2). Let us verify the uniqueness. If $\mathcal{A} \simeq Alg\,\mathcal{T}$ for some algebraic theory $\mathcal{T}$, then $\mathcal{A}_{pp} \simeq (Alg\,\mathcal{T})_{pp} \simeq Ic(\mathcal{T}^{op})$ by 6.14 and Corollary 8.9. Finally, $\mathcal{A}_{pp}^{op} \simeq Ic\,\mathcal{T}$ by Remark 8.5. □

8.13 Example

1. The canonical theory of the category *Set* is the theory $\mathcal{N}$ of natural numbers (see 1.4): it is clear that $\mathcal{N}$ is idempotent complete.
2. The canonical theory of the category *Ab* is the theory $\mathcal{T}_{ab}$ described in 1.20. In fact, we saw in 5.6 that $\mathcal{T}_{ab}$ is dual to Ab_{pp}.
3. In the category *Bool* of boolean algebras, we have the algebras $\mathcal{P}X$ of all subsets of a set X. The free algebras on n generators $\mathcal{PP}n$ form, for $n \in \mathbb{N}$, a strong generator. As noted in 6.10, the dual of this full subcategory of *Bool* is a theory for *Bool*. However, this is not the canonical theory. In fact, the canonical theory is the dual of the full subcategory of all algebras $\mathcal{P}n$ for $n \in \mathbb{N} \setminus \{0\}$, or equivalently, the category of finite nonempty sets and functions. Since each $n > 0$ is injective in the category of finite sets, it is a retract of 2^n. Thus $\mathcal{P}n$ is a retract of $\mathcal{PP}n$.

Historical remarks

The idempotent completion can already be found in Mitchell's (1965) monograph. Bunge (1966) presented it as well and used it for proving Corollary 8.10. She called it idempotent splitting closure; other names were used as well, for example, Cauchy completion or Karoubi envelope. Corollary 8.10 was later proved by Elkins and Zilber (1976). Proposition 8.12 is from Dukarm (1988).

9
Algebraic functors

We have studied algebraic categories as individual categories so far. It turns out that there is a natural concept of morphism between algebraic categories, which we call an *algebraic functor*, so that we obtain a 2-category of algebraic categories. We then prove a duality result: this 2-category is biequivalent to the 2-category of canonical algebraic theories. We first need to introduce a concept of morphism between algebraic theories – this is quite obvious:

9.1 Definition Let $\mathcal{T}_1$ and $\mathcal{T}_2$ be algebraic theories. A functor $M\colon \mathcal{T}_1 \to \mathcal{T}_2$ is called a *morphism of algebraic theories* if it preserves finite products.

9.2 Notation For a morphism of theories $M\colon \mathcal{T}_1 \to \mathcal{T}_2$, we denote by

$$\mathrm{Alg}\, M\colon \mathrm{Alg}\, \mathcal{T}_2 \to \mathrm{Alg}\, \mathcal{T}_1$$

the functor defined on objects $A\colon \mathcal{T}_2 \to \mathrm{Set}$ by $A \mapsto A \cdot M$.

9.3 Proposition Let $M\colon \mathcal{T}_1 \to \mathcal{T}_2$ be a morphism of algebraic theories.

1. $\mathrm{Alg}\, M\colon \mathrm{Alg}\, \mathcal{T}_2 \to \mathrm{Alg}\, \mathcal{T}_1$ preserves limits and sifted colimits.
2. $\mathrm{Alg}\, M$ has a left adjoint $M^*\colon \mathrm{Alg}\, \mathcal{T}_1 \to \mathrm{Alg}\, \mathcal{T}_2$, which is the essentially unique functor that preserves sifted colimits and makes the square

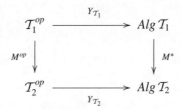

commutative up to natural isomorphism.

Proof The essentially unique functor M^* follows from Theorem 4.13. Since $Y_{\mathcal{T}_2} \cdot M^{op}$ preserves finite coproducts (see Lemma 1.13), to get the adjunction $M^* \dashv Alg\, M$, it suffices to apply 4.15: the right adjoint $Alg\, M$ is given by

$$B \mapsto Alg\, \mathcal{T}_2(Y_{\mathcal{T}_2}(M-), B).$$

By the Yoneda lemma, this is nothing but composition with M. This immediately implies that $Alg\, M$ preserves sifted colimits because they are calculated objectwise in $Alg\, \mathcal{T}_1$ and $Alg\, \mathcal{T}_2$ (see 2.5). □

9.4 Definition A functor between two algebraic categories is called *algebraic* provided that it preserves limits and sifted colimits.

9.5 Example

1. Every functor $Alg\, M$, for a theory morphism M, is algebraic.
2. The forgetful functor $Ab \to Set$ is algebraic.
3. Given an algebra A in an algebraic category $\mathcal{A}$, the hom-functor $\mathcal{A}(A, -)$: $\mathcal{A} \to Set$ is algebraic iff A is perfectly presentable.
4. A constant functor with value A between algebraic categories is algebraic iff A is a terminal object.
5. For every algebraic theory $\mathcal{T}$, the embedding $I: Alg\, \mathcal{T} \to Set^{\mathcal{T}}$ is an algebraic functor (see 1.21 and 2.5).

9.6 Remark We know from Proposition 9.3 that every morphism of theories induces an algebraic functor between the corresponding algebraic categories. If, moreover, the algebraic theories are canonical (8.11), then the algebraic functors are essentially just those induced by morphisms of theories (see Theorem 9.15). This will motivate us to define "morphisms of algebraic categories" as the algebraic functors. We are now going to prove that every algebraic functor has a left adjoint. For this we will use Freyd's adjoint functor theorem (see 0.8).

9.7 Theorem *A functor between algebraic categories is algebraic iff it has a left adjoint and preserves sifted colimits.*

Proof Let $G: \mathcal{B} \to \mathcal{A}$ be an algebraic functor. We are to prove that G has a left adjoint; that is, for every object A of $\mathcal{A}$, we are to prove that the functor

$$\mathcal{A}(A, G-): \mathcal{B} \to Set$$

is representable.

1. Assume first that A is perfectly presentable. Since G preserves limits, it remains to be proven that $\mathcal{A}(A, G-)$ satisfies the solution set condition of 0.8.

Every object B of $\mathcal{B}$ is a sifted colimit of objects from $\mathcal{B}_{pp}$ (see 6.9). Let us write $(\sigma_X: X \to B)$ for the colimit cocone. Since G preserves sifted colimits,

($G\sigma_X$: $GX \to GB$) is also a colimit cocone. As $\mathcal{A}(A, -)$ preserves sifted colimits, every morphism b: $A \to GB$ factorizes as follows:

2. If A is an arbitrary object of $\mathcal{A}$, we express it as a sifted colimit of perfectly presentable objects (see 6.9), say, $A = \text{colim}\, A_i$. From point 1 we know that $\mathcal{A}(A_i, G-)$ is representable, say, $\mathcal{A}(A_i, G-) \simeq \mathcal{B}(B_i, -)$. The Yoneda lemma now allows us to define an (obvious) sifted diagram whose objects are the B_i. Therefore $\mathcal{A}(A, G-)$ is representable by $\text{colim}\, B_i$ since

$$\mathcal{A}(A, G-) = \mathcal{A}(\text{colim}\, A_i, G-) \simeq \lim \mathcal{A}(A_i, G-)$$
$$\simeq \lim \mathcal{B}(B_i, -) \simeq \mathcal{B}(\text{colim}\, B_i, -).$$

□

9.8 Remark The previous theorem can be refined, as we demonstrate in Chapter 18: a functor between algebraic categories is algebraic iff it preserves limits, filtered colimits, and regular epimorphisms. One implication follows from the fact that (1) filtered implies sifted and (2) every regular epimorphism is a reflexive coequalizer (of its kernel pair). The converse implication is a particular case of 18.2.

9.9 Remark We are going to prove a duality between algebraic categories and canonical algebraic theories. This does not really mean a contravariant equivalence of categories. Indeed, a more subtle formulation is needed: just look at the simplest algebraic category, Set, and the simplest endomorphism, the identity functor Id_{Set}. It is easy to find a proper class of functors naturally isomorphic to Id_{Set} – and each of them is algebraic. However, in the category of all theories, no such phenomenon occurs. We thus need to work with morphisms of algebraic categories "up to natural isomorphism." For this reason, we have to move from categories to 2-categories. The reader does not need to know much about 2-categories. Here we summarize the needed facts.

9.10 A primer on 2-categories

1. Let us recall that a *2-category* $\mathbb{A}$ has a class $\text{obj}\,\mathbb{A}$ of objects, and instead of hom-sets $\mathbb{A}(A, B)$, it has hom-categories $\mathbb{A}(A, B)$ for every pair A, B of objects. The objects of $\mathbb{A}(A, B)$ are called *1-cells* and the morphisms

2-cells. Composition is represented by functors

$$c_{A,B,C}: \mathbb{A}(A, B) \times \mathbb{A}(B, C) \to \mathbb{A}(A, C),$$

which are associative in the expected sense: the two canonical functors

$$c_{A,B,D} \cdot (\mathbb{A}(A, B) \times c_{B,C,D}) \quad \text{and} \quad c_{A,C,D} \cdot (c_{A,B,C} \times \mathbb{A}(C, D))$$

from $\mathbb{A}(A, B) \times \mathbb{A}(B, C) \times \mathbb{A}(C, D)$ into $\mathbb{A}(A, D)$ are required to be equal. The identity morphisms are represented by functors

$$i_A: \mathbb{I} \to \mathbb{A}(A, A)$$

(where $\mathbb{I}$ is the 1-morphism category) which fulfill the usual requirements: both the composites

$$c_{A,A,B} \cdot (i_A \times \mathbb{A}(A, B)): \mathbb{I} \times \mathbb{A}(A, B) \to \mathbb{A}(A, B)$$
$$c_{A,B,B} \cdot (\mathbb{A}(A, B) \times i_B): \mathbb{A}(A, B) \times \mathbb{I} \to \mathbb{A}(A, B)$$

are the canonical isomorphisms.
2. A prototype of a 2-category is the 2-category *Cat* of all small categories (as objects). Given small categories A, B, then 1-cells are functors from A to B and 2-cells are natural transformations. In our book, we essentially work just with this 2-category and its sub-2-categories (see point 4).
3. Let us recall the concept of a *2-functor* $F: \mathbb{A} \to \mathbb{B}$ between 2-categories $\mathbb{A}$ and $\mathbb{B}$: it assigns objects FA of $\mathbb{B}$ to objects A of $\mathbb{A}$; for every pair A, A' of objects of $\mathbb{A}$, it defines a functor

$$F_{A,A'}: \mathbb{A}(A, A') \to \mathbb{B}(FA, FA').$$

Preservation of composition is expressed by the requirement that the two canonical functors

$$F_{A,C} \cdot c_{A,B,C} \quad \text{and} \quad c_{FA,FB,FC} \cdot (F_{A,B} \times F_{B,C})$$

from $\mathbb{A}(A, B) \times \mathbb{A}(B, C)$ into $\mathbb{B}(FA, FC)$ are equal. Preservation of identity morphisms is expressed by

$$F_{A,A} \cdot i_A = i_{FA}: \mathbb{I} \to \mathbb{B}(FA, FA).$$

4. A *sub-2-category* $\mathbb{A}$ of a 2-category $\mathbb{B}$ is given by a choice of a class of objects $\text{obj}\,\mathbb{A} \subseteq \text{obj}\,\mathbb{B}$ and, for every pair $A, A' \in \text{obj}\,\mathbb{A}$, of a subcategory $\mathbb{A}(A, A') \subseteq \mathbb{B}(A, A')$ such that composition and identity functors of $\mathbb{B}$ can be restricted to $\mathbb{A}$. The inclusion $I: \mathbb{A} \to \mathbb{B}$ is a 2-functor. The sub-2-category is *full* if $\mathbb{A}(A, A') = \mathbb{B}(A, A')$.

5. Two objects A, B of a 2-category $\mathbb{A}$ are called *equivalent* if there exist 1-cells $f\colon A \to B$ and $\overline{f}\colon B \to A$ such that $\overline{f} \cdot f$ is isomorphic to id_A in $\mathbb{A}(A, A)$ and $f \cdot \overline{f}$ is isomorphic to id_B in $\mathbb{A}(B, B)$.
6. A 2-functor $F\colon \mathbb{A} \to \mathbb{B}$ is called a *biequivalence* if all the functors $F_{A,A'}$ are equivalence functors, and every object of $\mathbb{B}$ is equivalent to FA for some object A of $\mathbb{A}$.
7. In contrast with equivalences of categories (see 0.2 and 0.3), if $F\colon \mathbb{A} \to \mathbb{B}$ is a biequivalence, then the quasi-inverses $F'\colon \mathbb{B} \to \mathbb{A}$ are no longer 2-functors but just homomorphisms of 2-categories as defined by Bénabou (1967). This means that the canonical requirements about compositions and identities are fulfilled by F' only up to invertible 2-cells.
8. For every 2-category $\mathbb{A}$, we denote by $\mathbb{A}^{op}$ the *dual 2-category*: it has the same objects, and the direction of 1-cells is reversed (while the direction of 2-cells remains nonreversed): $\mathbb{A}^{op}(A, A') = \mathbb{A}(A', A)$.

9.11 Definition

1. The *2-category Th of theories* has
    objects: algebraic theories
    1-cells: morphisms of algebraic theories
    2-cells: natural transformations
   This is a sub-2-category of *Cat*; that is, the composition of 1-cells and 2-cells are defined in *Th* as the usual composition of functors and natural transformations, respectively.
2. The *2-category Th $_c$ of canonical theories* is the full sub-2-category of *Th* on all theories that are canonical, that is, idempotent complete.
3. The *2-category ALG of algebraic categories* has
    objects: algebraic categories
    1-cells: algebraic functors
    2-cells: natural transformations
   Once again, composition is the usual composition of functors or natural transformations.

9.12 Remark

We need to be a little careful about foundations here: there is, as remarked earlier, a proper class of 1-cells in *ALG* (*Set*, *Set*), for example. However, if we consider the 1-cells up to natural isomorphism, all problems disappear: this is one consequence of the duality theorem 9.15. Ignoring the foundational considerations, we consider *ALG* as a sub-2-category of the 2-category of all categories. (The duality we prove subsequently tells us that *ALG* is essentially just the dual of *Th $_c$*.)

9.13 Definition We denote by

$$\mathrm{Alg}\colon \mathrm{Th}^{op} \to \mathrm{ALG}$$

the 2-functor assigning to every algebraic theory $\mathcal{T}$ the category $\mathrm{Alg}\,\mathcal{T}$, to every 1-cell $M\colon \mathcal{T}_1 \to \mathcal{T}_2$ the functor $\mathrm{Alg}\,M = (-) \cdot M$, and to every 2-cell $\alpha\colon M \to N$ the natural transformation $\mathrm{Alg}\,\alpha\colon \mathrm{Alg}\,M \to \mathrm{Alg}\,N$ whose component at a $\mathcal{T}_2$-algebra A is $A \cdot \alpha\colon A \cdot M \to A \cdot N$.

9.14 Remark The 2-functor Alg is well defined because of 9.3: for every morphism of theories M, the functor $\mathrm{Alg}\,M$ is algebraic. The fact that for every natural transformation α we get a natural transformation $\mathrm{Alg}\,\alpha$ is easy to verify.

9.15 Theorem: *Duality of algebraic categories and theories* The 2-category ALG of algebraic categories is biequivalent to the dual of the 2-category Th_c of canonical algebraic theories.

In fact, the domain restriction of the 2-functor Alg to canonical algebraic theories

$$\mathrm{Alg}\colon \mathrm{Th}_c^{op} \to \mathrm{ALG}$$

is a biequivalence.

Proof
1. Following 8.12, every algebraic category $\mathcal{A}$ is equivalent to $\mathrm{Alg}\,\mathcal{T}$ for the canonical algebraic theory $\mathcal{T} = \mathcal{A}_{pp}^{op}$.

2. We will prove that for two canonical algebraic theories $\mathcal{T}_1$ and $\mathcal{T}_2$, the functor

$$\mathrm{Alg}_{\mathcal{T}_1,\mathcal{T}_2}\colon \mathrm{Th}_c(\mathcal{T}_1, \mathcal{T}_2) \to \mathrm{ALG}(\mathrm{Alg}\,\mathcal{T}_2, \mathrm{Alg}\,\mathcal{T}_1)$$

is an equivalence of categories, see 0.3.

2a. $\mathrm{Alg}_{\mathcal{T}_1,\mathcal{T}_2}$ is full and faithful: given morphisms $M, N\colon \mathcal{T}_1 \rightrightarrows \mathcal{T}_2$ and a natural transformation $\lambda\colon \mathrm{Alg}\,M \to \mathrm{Alg}\,N$, there exists a unique natural transformation $\alpha\colon M \to N$ such that $\mathrm{Alg}\,\alpha = \lambda$. The proof follows the lines of the proof of the Yoneda lemma. Let us indicate how to construct α. Consider an object X in $\mathcal{T}_1$. Since $\mathcal{T}_2(MX, -) \in \mathrm{Alg}\,\mathcal{T}_2$, we have the component $\lambda_{\mathcal{T}_2(MX,-)}(X)\colon \mathcal{T}_2(MX, MX) \to \mathcal{T}_2(MX, NX)$, and we put

$$\alpha_X = \lambda_{\mathcal{T}_2(MX,-)}(X)(\mathrm{id}_{MX})\colon MX \to NX.$$

The family $(\alpha_X)_{X \in \mathcal{T}_1}$ is the required natural transformation $\alpha\colon M \to N$.

2b. The functor Alg_{T_1,T_2} is essentially surjective. In fact, consider an algebraic functor $G: Alg\, T_2 \to Alg\, T_1$, and let L be its left adjoint (see Theorem 9.7). We are going to prove that L can be restricted to a functor F that preserves finite coproducts and makes the square

$$\begin{array}{ccc} T_1^{op} & \xrightarrow{Y_{T_1}} & Alg\, T_1 \\ {\scriptstyle F}\downarrow & & \downarrow{\scriptstyle L} \\ T_2^{op} & \xrightarrow[Y_{T_2}]{} & Alg\, T_2 \end{array}$$

commutative up to natural isomorphism. For every object X in T_1, we have by adjunction a natural isomorphism

$$Alg\, T_2(L(Y_{T_1}(X)), -) \simeq Alg\, T_1(Y_{T_1}(X), G-).$$

Since $Y_{T_1}(X)$ is perfectly presentable (see 5.5) and G preserves sifted colimits, the natural isomorphism above implies that $L(Y_{T_1}(X))$ is perfectly presentable. By 5.14, $L(Y_{T_1}(X))$ is a retract of a representable algebra, and since T_2 is idempotent complete, $L(Y_{T_1}(X))$ is itself a representable algebra (see 8.3.3). This means that there exists an essentially unique object FX in T_2 such that $L(Y_{T_1}(X)) \simeq Y_{T_2}(FX)$. In this way, we get a map on objects, $F: \operatorname{obj} T_1 \to \operatorname{obj} T_2$, that, by the Yoneda lemma, extends to a functor $F: T_1^{op} \to T_2^{op}$, making the preceding square commutative up to natural isomorphism. F preserves finite coproducts because Y_{T_1} preserves them by 1.13, Y_{T_2} reflects finite coproducts, and L preserves them. It remains to be proven that $G \simeq Alg\, M$, where $M = F^{op}: T_1 \to T_2$, or equivalently, that $L \simeq M^*$ in the notation of Proposition 9.3. But this follows from the essential commutativity of the preceding square and the last part of Proposition 9.3. This proves that Alg_{T_1,T_2} is essentially surjective. $\square$

9.16 Corollary *A functor between algebraic categories*

$$G: \mathcal{A}_2 \to \mathcal{A}_1$$

is algebraic iff it is induced by a morphism of theories; that is, there exists a morphism of the corresponding canonical algebraic theories $M: T_1 \to T_2$ *and two equivalence functors* $E_1: Alg\, T_1 \to \mathcal{A}_1$ *and* $E_2: Alg\, T_2 \to \mathcal{A}_2$ *such that*

the square

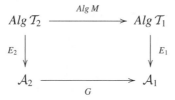

commutes up to natural isomorphism.

In fact, given G, the proof of Theorem 9.15 yields $G \simeq Alg\, M$, with the desired property. The converse implication is clear.

9.17 Remark The category Set is a kind of *dualizing object* for the biequivalence $Alg\colon Th_c^{op} \to ALG$:

1. Forgetting size considerations (an algebraic theory is by definition a small category), we have $Alg\, \mathcal{T} = Th(\mathcal{T}, Set)$ for every algebraic theory $\mathcal{T}$.
2. For every algebraic category $\mathcal{A}$, there is an equivalence of categories

$$\mathcal{A}_{pp}^{op} \simeq ALG(\mathcal{A}, Set).$$

Indeed, if $A \in \mathcal{A}_{pp}^{op}$, then $G = \mathcal{A}(A, -)\colon \mathcal{A} \to Set$ preserves limits and sifted colimits. Conversely, let $G\colon \mathcal{A} \to Set$ be an algebraic functor and let L be a left adjoint of G (see Theorem 9.7). Then $G \simeq \mathcal{A}(L1, -)$ (with 1 denoting a one-element set), and $L1$ is perfectly presentable because G preserves sifted colimits.

9.18 Remark We conclude this chapter by mentioning the analogous Gabriel–Ulmer duality for locally finitely presentable categories. The proof is similar to that of Theorem 9.15. Whereas the morphisms of algebraic categories are the functors preserving limits and sifted colimits, the morphisms of locally finitely presentable categories are the functors preserving limits and filtered colimits. These are the 1-cells of the 2-category LFP, and the 2-cells are natural transformations. We also denote by LEX the 2-category of small categories with finite limits, functors preserving finite limits, and natural transformations. The 2-functor

$$Lex\colon LEX^{op} \to LFP$$

assigns to every small category $\mathcal{T}$ with finite limits the category $Lex\, \mathcal{T}$ (4.16), and it acts on 1-cells and 2-cells in an analogous way to $Alg\colon Th^{op} \to ALG$.

9.19 Theorem *The 2-categories LFP and LEX are dually biequivalent. In fact, Lex: $LEX^{op} \to LFP$ is a biequivalence. The converse construction associates with a locally finitely presentable category $\mathcal{A}$ the small, finitely complete category $\mathcal{A}_{fp}^{op}$ (6.26).*

Historical remarks

Morphisms of algebraic theories and the resulting algebraic functors were (in the one-sorted case) introduced by Lawvere (1963) and belong to the main contribution of his work. The characterization 9.7 of algebraic functors and the duality theorem 9.15 are contained in Adámek et al. (2003). This is analogous to the Gabriel–Ulmer duality of 9.19 for locally finitely presentable categories (see Gabriel & Ulmer, 1971). A general result can be found in Centazzo and Vitale (2002). 2-categories have been introduced by Ehresmann (1963).

10
Birkhoff's variety theorem

So far we have not treated one of the central concepts of algebra: equations. In the present chapter, we prove the famous characterization of varieties of algebras, due to G. Birkhoff: varieties are precisely the full subcategories of $Alg\,T$ closed under

> products
> subalgebras
> regular quotients
> directed unions

The last item was not included in Birkhoff's formulation. The reason is that Birkhoff only considered one-sorted algebras, and for them, directed unions follow from the other three items (see 11.34). For general algebraic categories, directed unions cannot be omitted (see Example 10.23).

Classically, an equation is an expression $u = v$ where u and v are terms (say, in n variables). We will see in 13.9 that such terms are morphisms from n to 1 in the theory of Σ-algebras. We can also consider k-tuples of classical equations as pairs of morphisms from n to k. The following concept generalizes this idea.

10.1 Definition If T is an algebraic theory, an *equation* in T is a parallel pair $u, v\colon s \rightrightarrows t$ of morphisms in T. (Following algebraic tradition, we write $u = v$ in place of (u, v).) An algebra $A\colon T \to Set$ *satisfies* the equation $u = v$ if $A(u) = A(v)$.

10.2 Example

1. In the theory T_{ab} of abelian groups (1.6), recall that endomorphisms of 1 have the form $[n]$ and correspond to the operations on abelian groups given

by $x \mapsto n \cdot x$. Thus the equation

$$[2] = [0]$$

is satisfied by precisely the groups with

$$x + x = 0$$

for all elements x.

2. Graphs whose only edges are loops are given, considering the theory in 1.11, by the equation

$$\tau = \sigma.$$

10.3 Remark Observe that if an algebra A satisfies the equation $u = v$, then it also satisfies all the equations of the form $u \cdot x = v \cdot x$ and $y \cdot u = y \cdot v$ for $x: s' \to s$ and $y: t \to t'$ in $\mathcal{T}$. Moreover, if the equations $u_i = v_i$, $i = 1, \ldots, n$, are satisfied by A for $u_i, v_i: s \to t_i$, then A also satisfies $\langle u_i \rangle = \langle v_i \rangle$, where

$$\langle u_i \rangle, \langle v_i \rangle: s \to t_1 \times \ldots \times t_n$$

are the corresponding morphisms. For this reason, we will state the definition of variety using congruences as well as equations.

10.4 Definition Let $\mathcal{T}$ be an algebraic theory. A *congruence* on $\mathcal{T}$ is a collection $\sim$ of equivalence relations $\sim_{s,t}$ on hom-sets $\mathcal{T}(s, t)$, where (s, t) ranges over pairs of objecs, which is stable under composition and finite products in the following sense:

1. If $u \sim_{s,t} v$ and $x \sim_{r,s} y$, then $u \cdot x \sim_{r,t} v \cdot y$:

$$r \underset{y}{\overset{x}{\rightrightarrows}} s \underset{v}{\overset{u}{\rightrightarrows}} t.$$

2. If $u_i \sim_{s,t_i} v_i$ for $i = 1, \ldots, n$, then $\langle u_1, \ldots, u_n \rangle \sim_{s,t} \langle v_1, \ldots, v_n \rangle$, where $t = t_1 \times \ldots \times t_n$:

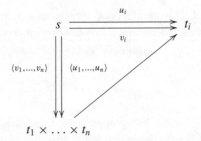

10.5 Example Consider the theory $\mathcal{T}_{mon}$ of monoids (3.10) whose morphisms from n to k are all k-tuples of words over $n = \{0, 1, \ldots, n-1\}$, see 11.23. The commutative law corresponds to the congruence $\sim$ with $\sim_{n,1}$ defined for words v and w by

$w \sim_{n,1} v$ iff v can be obtained from w by a permutation of its letters.

The general equivalence $\sim_{n,k}$ is defined by coordinates of the k-tuples of words.

10.6 Example Let $\mathcal{T}$ be an algebraic theory and let $M: \mathcal{T} \to \mathcal{B}$ be a finite product-preserving functor. The *kernel congruence* of M is denoted by $\approx_M$; that is, for all $u, v \in \mathcal{T}(s, t)$, we put

$$u \approx_M v \quad \text{iff} \quad Mu = Mv.$$

This is obviously a congruence on $\mathcal{T}$. In particular, for every $\mathcal{T}$-algebra A, we have a congruence $\approx_A$.

10.7 Remark

1. Congruences on a given algebraic theory $\mathcal{T}$ are ordered in a canonical way: we write $\sim \,\subseteq\, \sim'$ in the case that for every pair s, t of objects of $\mathcal{T}$, we have

$$u \sim_{s,t} v \quad \text{implies} \quad u \sim'_{s,t} v.$$

2. It is easy to see that every (set theoretical) intersection of congruences is a congruence. Consequently, for every set E of equations, there exists the smallest congruence $\sim_E$ on $\mathcal{T}$ containing E. We say that the congruence $\sim_E$ is *generated* by the equations of E.

10.8 Definition A full subcategory $\mathcal{A}$ of $Alg\,\mathcal{T}$ is called a *variety* if there exists a set of equations such that a $\mathcal{T}$-algebra lies in $\mathcal{A}$ iff it satisfies all equations in that set.

10.9 Remark

1. Varieties are also sometimes called *equational classes of algebras* or *equational categories*. But we reserve this name for the special case of varieties of Σ-algebras treated in Chapters 13 (for one-sorted signatures) and 14 (for S-sorted ones).
2. We will use the name *variety* also in the loser sense of a category *equivalent* to a full subcategory of $Alg\,\mathcal{T}$ specified by equations. Every time we use the word *variety*, it will be clear whether the preceding definition or the loser version is meant.

10.10 Example

1. The abelian groups satisfying $x + x = 0$ form a variety in *Ab*.
2. All graphs whose edges are just loops form a variety in *Graph*.
3. Let us consider the category *Set* × *Set* of two-sorted sets with sorts called, say, s and t. This has an algebraic theory $\mathcal{T}_C$ of all words over $\{s, t\}$ (see 1.5). Consider the full subcategory $\mathcal{A}$ of *Set* × *Set* of all pairs $A = (A_s, A_t)$ with either $A_s = \emptyset$ or A_t having at most one element. This can be specified by the equation given by the parallel pair of projections

$$stt \rightrightarrows t.$$

10.11 Remark Every variety is also specified by a congruence. More precisely, given a variety $\mathcal{A}$ of $Alg\,\mathcal{T}$, let $\sim$ be the congruence generated by the given set E of equations. A $\mathcal{T}$-algebra A lies in $\mathcal{A}$ iff it satisfies all equations in $\sim$, that is, iff it fulfills

$$u \sim v \quad \text{implies} \quad Au = Av.$$

In fact, if A satisfies all equations in E, then $E \subseteq \approx_A$ and therefore $\sim\, \subseteq\, \approx_A$.

10.12 Notation For every congruence $\sim$ on an algebraic theory $\mathcal{T}$, we denote by

$$\mathcal{T}/\sim$$

the algebraic theory on the same objects and with morphisms given by the congruence classes of morphisms of $\mathcal{T}$:

$$(\mathcal{T}/\sim)(s, t) = \mathcal{T}(s, t)/\sim_{s,t}$$

Composition and identity morphisms are inherited from $\mathcal{T}$; more precisely, they are determined by the fact that we have a functor

$$Q: \mathcal{T} \to \mathcal{T}/\sim$$

which is the identity map on objects and which assigns to every morphism its congruence class.

10.13 Remark

1. It is easy to verify that $\mathcal{T}/\sim$ has finite products determined by those of $\mathcal{T}$; thus $\mathcal{T}/\sim$ is an algebraic theory and Q is a theory morphism. Moreover, the functor Q is full and surjective on objects.

2. A morphism of theories $M: T \to T'$ factorizes through Q up to natural isomorphism

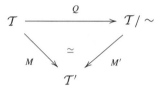

iff the congruence $\sim$ is contained in the kernel congruence $\approx_M$. When this is the case, the factorization M' is essentially unique and is a theory morphism. In fact, it is clear that if $M' \cdot Q \simeq M$, then $\sim \; \subseteq \; \approx_M$. For the converse, define M' to be equal to M on objects, and put $M'[f] = Mf$ on morphisms. Clearly $M' \cdot Q = M$. From the fact that M preserves finite products and Q reflects them, the desired properties of M' easily follow.

3. If in point 2 we take $\sim$ to be equal to the congruence $\approx_M$, then the factorization M',

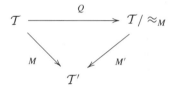

is faithful. Therefore M is full and essentially surjective iff M' is an equivalence of categories.

10.14 Proposition *Let $\sim$ be a congruence on an algebraic theory T, and let $Q: T \to T/\sim$ be the corresponding quotient. The functor*

$$\mathrm{Alg}\, Q: \mathrm{Alg}\,(T/\sim) \to \mathrm{Alg}\, T$$

is full and faithful and injective on objects.

Proof

1. Clearly $\mathrm{Alg}\, Q$ is faithful because Q is surjective.

2. Consider objects $A, B \in \mathrm{Alg}\, T/\sim$ and a morphism $\beta: A \cdot Q \to B \cdot Q$, that is, a collection $\beta_t: At \to Bt$ of homomorphisms natural in t ranging through T. Then the same collection is natural in t ranging through $T/\sim$ and thus $\beta: A \to B$ is a morphism of $\mathrm{Alg}\,(T/\sim)$. Clearly $\mathrm{Alg}\, Q$ takes this morphism to the original one. □

10.15 Corollary *Every variety is an algebraic category.*

In more detail, let $\sim$ be a congruence on an algebraic theory $\mathcal{T}$, and let $\mathcal{A}$ be the full subcategory of $Alg\,\mathcal{T}$ specified by $\sim$. There exists an isomorphism of categories
$$E: Alg\,(\mathcal{T}/\sim) \to \mathcal{A}$$
such that the triangle

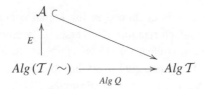

commutes.

In fact, for every $(\mathcal{T}/\sim)$-algebra B, given $u \sim v$, we have $B(Qu) = B[u] = B[v] = B(Qv)$. This implies that $Alg\,Q$ factorizes through the inclusion of $\mathcal{A}$ in $Alg\,\mathcal{T}$. Moreover, if A lies in $\mathcal{A}$, then $A = (Alg\,Q)(B)$, where B is the $(\mathcal{T}/\sim)$-algebra defined by $B[u] = Au$. This shows that the factorization $E: Alg\,(\mathcal{T}/\sim) \to \mathcal{A}$ is bijective on objects. Since $Alg\,Q$ is full and faithful (see Proposition 10.14), E is an isomorphism.

10.16 Proposition *Every variety $\mathcal{A}$ of $\mathcal{T}$-algebras is closed in $Alg\,\mathcal{T}$ under*

1. *products: given a product $B = \prod_{i \in I} A_i$ in $Alg\,\mathcal{T}$ with all A_i in $\mathcal{A}$, B also lies in $\mathcal{A}$*
2. *subalgebras: given a monomorphism $m: B \to A$ in $Alg\,\mathcal{T}$ with A in $\mathcal{A}$, B also lies in $\mathcal{A}$*
3. *regular quotients: given a regular epimorphism $e: A \to B$ in $Alg\,\mathcal{T}$ with A in $\mathcal{A}$, B also lies in $\mathcal{A}$*
4. *sifted colimits: given a sifted colimit $B = \mathrm{colim}\,A_i$ in $Alg\,\mathcal{T}$ with all A_i in $\mathcal{A}$, B also lies in $\mathcal{A}$*

Proof Following Corollary 10.15, the inclusion functor $\mathcal{A} \to Alg\,\mathcal{T}$ is naturally isomorphic to $Alg\,Q$, which preserves limits and sifted colimits by 9.3. This proves points 1 and 4.

Let $m: B \to A$ be a monomorphism with A in $\mathcal{A}$. We prove that B is in $\mathcal{A}$ by verifying that every equation $u_1, u_2: s \rightrightarrows t$ that A satisfies is also satisfied by B. We know from 1.23.2 that the component $m_t: Bt \to At$ is a monomorphism.

From $A(u_1) = A(u_2)$ and the commutativity of the squares,

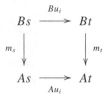

we conclude that $Bu_1 = Bu_2$.

Let $e: A \to B$ be a regular epimorphism with A in $\mathcal{A}$. To prove that B lies in $\mathcal{A}$, observe that a kernel pair $k_1, k_2: N(e) \rightrightarrows A$ of e yields a subobject of $A \times A \in \mathcal{A}$, and thus $N(e)$ lies in $\mathcal{A}$ by points 1 and 2. And since the pair k_1, k_2 is reflexive and e is its coequalizer, B is a sifted colimit (3.2) of a diagram in $\mathcal{A}$, and thus $B \in \mathcal{A}$. □

10.17 Corollary *Every variety $\mathcal{A}$ of $\mathcal{T}$-algebras is closed in Alg $\mathcal{T}$ under limits and sifted colimits.*

10.18 Example Not every full subcategory of an algebraic category $Alg\,\mathcal{T}$ closed under limits and sifted colimits is a variety. Consider the free completion $\mathcal{T}'$ of $\mathcal{T}$ under finite products (1.14) and the finite product–preserving extension $M: \mathcal{T}' \to \mathcal{T}$ of the identity functor on $\mathcal{T}$. The induced functor $Alg\,M$ is naturally isomorphic to the full inclusion $Alg\,\mathcal{T} \to Alg\,\mathcal{T}' \simeq Set^{\mathcal{T}}$, but $Alg\,\mathcal{T}$ is usually not closed in $Set^{\mathcal{T}}$ under subalgebras (in contrast to Proposition 10.16). As a concrete example, let $\mathcal{T}$ be the category of finite sets and functions. The functor

$$A: \mathcal{T} \to Set, \quad AX = X \times X \times X$$

is clearly an algebra for $\mathcal{T}$, but its subfunctor A' given by all triples in $X \times X \times X$ in which at least two different coordinates have the same value is not an algebra for $\mathcal{T}$.

10.19 Remark Following Remark 10.13.3 and Proposition 10.14, the algebraic functor induced by a full and essentially surjective morphism of theories is full and faithful. The morphism $M: \mathcal{T}' \to \mathcal{T}$ in Example 10.18 also demonstrates that the converse implication does not hold: $Alg\,M: Alg\,\mathcal{T} \to Set^{\mathcal{T}}$ is full and faithful, but M is not full.

10.20 Definition A full reflective subcategory $\mathcal{A}$ of a category $\mathcal{B}$ (see 0.9) is called *regular epireflective* if all reflections $r_B: B \to RB$ are regular epimorphisms.

10.21 Corollary *Every variety of T-algebras is a regular epireflective subcategory of $\text{Alg}\,T$ closed under regular quotients and directed unions.*

Proof We already know that the variety $\mathcal{A}$ is closed in $\text{Alg}\,T$ under regular quotients and sifted colimits (Proposition 10.16) and therefore under directed unions, which are a special case of sifted colimits (2.9). Moreover, following 10.15, the inclusion functor $\mathcal{A} \to \text{Alg}\,T$ is naturally isomorphic to $\text{Alg}\,Q$, which has a left adjoint (9.3). It remains to prove that for every T-algebra B, the reflection $r_B \colon B \to RB$ is a regular epimorphism. Let $r_B = m \cdot e$

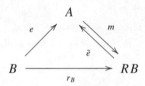

be a regular factorization of r_B (see 3.7). By Proposition 10.16.2, $A \in \mathcal{A}$, and thus there is a unique $\bar{e} \colon RB \to A$ such that $e = \bar{e} \cdot r_B$. Since e is an epimorphism, we see that $\bar{e} \cdot m = \text{id}_A$. Also $m \cdot \bar{e} = \text{id}_{RB}$ due to the universal property of r_B. Thus m is an isomorphism and r_B a regular epimorphism. □

10.22 Birkhoff's variety theorem *Let T be an algebraic theory. A full subcategory $\mathcal{A}$ of $\text{Alg}\,T$ is a variety iff it is closed in $\text{Alg}\,T$ under*

1. *products*
2. *subalgebras*
3. *regular quotients and*
4. *directed unions*

Proof Every variety is closed under points 1–4 (see Proposition 10.16). Conversely, let $\mathcal{A}$ be closed under points 1–4.

1. We first prove that $\mathcal{A}$ is a regular epireflective subcategory. Let B be a T-algebra. By 3.6, there exists a set of regular epimorphisms $e \colon B \to A_e$ ($e \in X$) representing all regular quotients of B with codomains in $\mathcal{A}$. Denote by $b \colon B \to \prod_{e \in X} A_e$ the induced morphism, and let

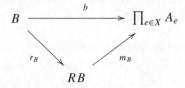

be the regular factorization of b (see 3.7). The algebra RB lies in $\mathcal{A}$ (because it is a subalgebra of a product of algebras in $\mathcal{A}$), and r_B is the reflection of B in $\mathcal{A}$. Indeed, for every morphism $f: B \to A$ with A in $\mathcal{A}$, we have a regular factorization

$$f = m \cdot e \quad \text{for some} \quad e \in X,$$

and since e factorizes through b, so does f.

2. We will prove that $\mathcal{A}$ is specified by the congruence $\sim$ which is the intersection of the kernel congruences of all algebras in $\mathcal{A}$:

$$u_1 \sim_{s,t} u_2 \quad \text{iff} \quad Au_1 = Au_2 \quad \text{for all} \quad A \in \mathcal{A}.$$

This is indeed a congruence (see Example 10.6 and Remark 10.7.2). It is our task to prove that every $\mathcal{T}$-algebra B such that $\sim$ is contained in $\approx_B$ (see Remark 10.7), lies in $\mathcal{A}$.

2a. Assume first that B is a regular quotient of a representable algebra $Y_\mathcal{T}(t)$. We thus have a regular epimorphism $e: Y_\mathcal{T}(t) \to B$. We know that the reflection morphism $r_t: Y_\mathcal{T}(t) \to R(Y_\mathcal{T}(t))$ is a regular epimorphism, and thus it is a coequalizer:

$$N \underset{v}{\overset{u}{\rightrightarrows}} Y_\mathcal{T}(t) \xrightarrow{r_t} R(Y_\mathcal{T}(t)).$$

By 4.2, we can express N as a sifted colimit of representable algebras and denote the colimit cocone by $c_s: Y_\mathcal{T}(s) \to N$. Using the Yoneda lemma, we see that for every s, there exist morphisms $u_s, v_s: t \rightrightarrows s$ representing $u \cdot c_s$ and $v \cdot c_s$, respectively:

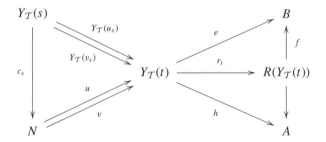

Moreover, we have $u_s \sim_{t,s} v_s$ because every morphism $h: Y_\mathcal{T}(t) \to A$ with $A \in \mathcal{A}$ factorizes through r_t so that $Y_\mathcal{T}(u_s) \cdot h = Y_\mathcal{T}(v_s) \cdot h$, and it follows that $Au_s = Av_s$. By assumption on B, this implies $Bu_s = Bv_s$, and then $e \cdot u \cdot c_s = e \cdot v \cdot c_s$. The cocone c_s is jointly epimorphic, and thus we have $e \cdot u = e \cdot v$. Since r_t is the coequalizer of u and v, there exists $f: R(Y_\mathcal{T}(t)) \to B$ such that

$f \cdot r_t = e$. Finally, f is a regular epimorphism because e is, so that B, being a regular quotient of $R(Y_T(t))$, lies in $\mathcal{A}$.

2b. Let B be arbitrary. Express B as a sifted colimit of representable algebras (4.2), and for each of the colimit morphisms $\sigma_s \colon Y_T(s) \to B$, denote by B_s the image of σ_s, which, by 3.7, is a subalgebra of B. Since T has finite products, the collection of these subalgebras of B is directed. Since $\mathcal{A}$ is closed under directed unions, we only need to prove that every B_s lies in $\mathcal{A}$. This follows from 2a: we know that B_s is a regular quotient of a representable algebra, and B_s has the desired property: given $u_1 \sim_{s,t} u_2$, we have $B(u_1) = B(u_2)$, and this implies $B_s u_1 = B_s u_2$ since B_s is a subalgebra of B. □

10.23 Example The assumption of closure under directed unions cannot be omitted: consider the category $Set^{\mathbb{N}}$ of $\mathbb{N}$-sorted sets, and let $\mathcal{A}$ be the full subcategory of all $A = (A_n)_{n \in \mathbb{N}}$ such that either some component A_n is empty or all components have precisely one element. This subcategory is clearly closed under products, subalgebras, and regular quotients – we omit the easy verification. However, it is not a variety, not being closed under directed unions. In fact, every $\mathbb{N}$-sorted set $A = (A_n)_{n \in \mathbb{N}}$ is a directed union of objects A^k of $\mathcal{A}$: put $A^k = (A_0, \ldots, A_k, \emptyset, \emptyset, \ldots)$.

10.24 Corollary *Let $\mathcal{A}$ be a full subcategory of $Alg\,T$. Then $\mathcal{A}$ is a variety iff it is a regular epireflective subcategory closed under regular quotients and directed unions.*

In fact, following Corollary 10.21 and Theorem 10.22, we only need to observe that every regular epireflective subcategory $\mathcal{A}$ is closed in $Alg\,T$ under products (and this is obvious) and subalgebras: consider the diagram

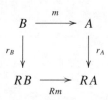

where m is a monomorphism and A lies in $\mathcal{A}$. Then r_A is an isomorphism, and r_B is a monomorphism. But r_B is also a regular epimorphism so that it is an isomorphism and B lies in $\mathcal{A}$.

10.25 Example The category Ab_{tf} of torsion-free abelian groups is a regular epireflective subcategory of Ab closed under filtered colimits. But this is not a variety in Ab because it is not closed under quotients. Indeed, Ab_{tf} is locally finitely presentable but not algebraic (it is not exact).

Historical remarks

The classical characterization of varieties of one-sorted algebras as HSP classes is from Birkhoff (1935). We present this version in 11.34.

For many-sorted algebras, the concept of equation in Definition 10.1 corresponds to the formulas

$$\forall x_1 \ldots \forall x_n \, (t = s),$$

where t and s are terms of the same sort in (sorted) variables $x_1, \ldots, x_n$. Example 10.23 demonstrates that with respect to the equations above, we need to add the closure under directed unions. Another approach is to admit infinitely many variables in the equations – that is, to work in the logic $L_{\omega\infty}$ (admitting quantification over infinite sets) rather than in the finitary logic $L_{\omega\omega}$. With this logic, directed unions can be omitted in Birkhoff's variety theorem. This is illustrated by Example 14.21 (see also J. Adámek et al., 2010).

Dropping in Corollary 10.24 the assumption of closure under regular quotients, one gets the so-called *quasivarieties*. Theories of quasivarieties were studied by Adámek and Porst (1998). A survey on quasivarieties based on free completions is presented by Pedicchio and Vitale (2000).

PART II

Concrete algebraic categories

The most elegant treatment of clones is given by F. W. Lawvere and J. Bénabou using categories.
— G. Grätzer (2008), *Universal Algebra,* Springer: 46

11
One-sorted algebraic categories

Classical algebraic categories, such as groups, modules, and boolean algebras, are not only abstract categories: their objects are sets with a structure, and their morphisms are functions preserving the structure. Thus they are concrete categories over *Set* (see 0.18), which means that a "forgetful" functor into *Set* is given. Also, these classical algebraic categories have theories generated by a single object X in the sense that all objects of the theory are finite powers X^n of X. (Consider the free group on one generator X: in the theory of groups, which is the dual of the category of finitely generated free groups, every object is a power of X.) To formalize this idea, we study in this chapter one-sorted algebraic categories. In Chapter 14, we will deal with the more general notion of S-sorted algebraic categories, for which the forgetful functor into a power of *Set* is considered rather than into *Set*.

11.1 Example The theory

$$\mathcal{N}$$

of sets (see 1.4), which is the full subcategory of Set^{op} on natural numbers $n = \{0, \ldots, n-1\}$, is a "prototype" one-sorted theory: every object n is the product of n copies of 1. Moreover, the n injections in *Set*,

$$\pi_i^n: 1 \to n, \quad 0 \mapsto i \quad (i = 0, \ldots, n-1),$$

yield a canonical choice of projections $\pi_i^n: n \to 1$ in $\mathcal{N}$ that present n as 1^n.

11.2 Remark

1. If an algebraic theory $\mathcal{T}$ has all objects given by finite powers of an object X, we obtain a theory morphism $T: \mathcal{N} \to \mathcal{T}$ as follows: for every n, choose an

nth power of X with projections $p_i^n\colon X^n \to X$ for $i = 0, \ldots, n-1$. Then T is uniquely determined by

$$Tn = X^n \quad \text{and} \quad T\pi_i^n = p_i^n \quad \text{for all } i < n \text{ in } \mathbb{N}.$$

2. Conversely, every theory morphism $T\colon \mathcal{N} \to \mathcal{T}$ represents an object and a choice of projections of all its finite powers: put $X = T1$ and $p_i^n = T\pi_i^n\colon X^n \to X$ for all $i < n$.

This leads us to the following

11.3 Definition

1. A *one-sorted algebraic theory* is a pair $(\mathcal{T}, T)$ in which $\mathcal{T}$ is an algebraic theory whose objects are the natural numbers, and $T\colon \mathcal{N} \to \mathcal{T}$ is a theory morphism that is the identity map on objects.
2. A *morphism* of one-sorted algebraic theories $M\colon (\mathcal{T}_1, T_1) \to (\mathcal{T}_2, T_2)$ is a functor $M\colon \mathcal{T}_1 \to \mathcal{T}_2$ such that $M \cdot T_1 = T_2$:

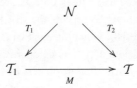

11.4 Remark

1. We have not requested that morphisms of one-sorted theories preserve finite products. In fact, this simply follows from the equation $M \cdot T_1 = T_2$. Observe that because of that equation M is the identity map on objects. Since, moreover, M takes the projections $T_1\pi_i^n$ to the projections $T_2\pi_i^n$, it clearly preserves finite powers and thus finite products.
2. The reason why one-sorted theories are requested to be equipped with a theory morphism from $\mathcal{N}$ is that in this way, the category of one-sorted theories and the category of finitary monads on *Set* are equivalent, as proven in Theorem A.37 (see Appendix A). (And by the way, this is the original definition by Lawvere from 1963.)
3. There is an obvious nonstrict version of morphism of one-sorted theories, where the equality $M \cdot T_1 = T_2$ is weakened to a natural isomorphism between $M \cdot T_1$ and T_2. See Appendix C for this approach.

11.5 Example Recall the theory $\mathcal{T}_{ab}$ of abelian groups whose objects are natural numbers and morphisms are matrices (see 1.6). It can be canonically considered as a one-sorted theory if we define $T_{ab}\colon \mathcal{N} \to \mathcal{T}_{ab}$ as the identity map on objects and assign to $\pi_i^n\colon n \to 1$ the one-row matrix with ith entry 1 and all other entries 0.

11.6 Remark Given a one-sorted theory $(\mathcal{T}, T)$, the functor T does not influence the concept of algebra: the category $Alg\,\mathcal{T}$ thus consists, again, of all functors $A\colon \mathcal{T} \to Set$ preserving finite products. However, the presence of T makes the category of algebras concrete over Set (see 0.18). Assuming that we identify Set and $Alg\,\mathcal{N}$, the forgetful functor is simply

$$Alg\,T\colon Alg\,\mathcal{T} \to Set$$

(see 9.2), which is faithful by Proposition 11.8. More precisely, this forgetful functor takes an algebra $A\colon \mathcal{T} \to Set$ to the set $A1$ and a homomorphism $h\colon A \to B$ to the component $h_1\colon A1 \to B1$.

11.7 Example For the one-sorted theory $(\mathcal{T}_{ab}, T_{ab})$ of abelian groups, the category $Alg\,\mathcal{T}_{ab}$ is equivalent to Ab. (But it is not isomorphic to Ab: this is caused by the fact that algebras for $\mathcal{T}_{ab}$ are not required to preserve products strictly. Consequently, there exist many algebras that are naturally isomorphic to algebras of the form $\widehat{G}$ (see 1.6) but are not equal to any of those.) The forgetful functor assigns to $\widehat{G}$, for every group G, the underlying set of G. Observe that unlike in Ab, the category $Alg\,\mathcal{T}_{ab}$ has the property that there exist isomorphisms $f\colon A \to B$ in $Alg\,\mathcal{T}_{ab}$ for which $A1 = B1$ and $f_1 = \mathrm{id}$ but still $A \neq B$. In fact, given a group G, we usually have many algebras $B \neq \widehat{G}$ naturally isomorphic to $\widehat{G}$ such that the component of the natural isomorphism at 1 is the identity. In other words, $Alg\,\mathcal{T}_{ab}$ is not amnestic (see 13.16).

11.8 Proposition *Let $(\mathcal{T}, T)$ be a one-sorted algebraic theory. The forgetful functor $Alg\,T\colon Alg\,\mathcal{T} \to Set$ is algebraic, faithful, and conservative.*

Proof $Alg\,T$ is algebraic by 9.3. Let $f\colon A \to B$ be a homomorphism of $\mathcal{T}$-algebras. Because of the naturality of f, the following square commutes:

$$\begin{array}{ccc} An & \xrightarrow{f_n} & Bn \\ \simeq\downarrow & & \downarrow\simeq \\ (A1)^n & \xrightarrow{(f_1)^n} & (B1)^n \end{array}$$

It is now obvious that $Alg\,T$ is faithful and conservative. □

11.9 Corollary *Let $(\mathcal{T}, T)$ be a one-sorted algebraic theory. The forgetful functor preserves and reflects limits, sifted colimits, monomorphisms, and regular epimorphisms.*

11.10 Remark For every $\mathcal{T}$-algebra A and every subset X of its underlying set $A1$, there exists the least subalgebra $\overline{X}$ of A such that $\overline{X}1$ contains X ($\overline{X}$ is called the *subalgebra generated* by X). In fact, consider the intersection of all subalgebras of A containing X.

11.11 Remark The concept of *algebraic category* in Chapter 1 used equivalences of categories. For one-sorted algebraic theories, we need more: the equivalence functor must be concrete. This is, in fact, not enough because the quasi-inverse (0.2.4) of a concrete functor is, in general, not concrete. We are going to require that the equivalence functor admits a quasi-inverse that is concrete:

11.12 Definition Given concrete categories $U: \mathcal{A} \to \mathcal{K}$ and $V: \mathcal{B} \to \mathcal{K}$ over $\mathcal{K}$, by a *concrete equivalence* between them we mean a pair of concrete functors

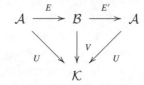

such that both $E \cdot E'$ and $E' \cdot E$ are naturally isomorphic to the identity functors. We then say that $(\mathcal{A}, U)$ and $(\mathcal{B}, V)$ are *concretely equivalent*.

11.13 Definition A *one-sorted algebraic category* is a concrete category over *Set* that is concretely equivalent to $Alg\,\mathcal{T}: Alg\,\mathcal{T} \to Set$ for a one-sorted algebraic theory $(\mathcal{T}, T)$.

11.14 Remark A nonstrict version of one-sorted algebraic categories is treated in Appendix C.

11.15 Example The category *Ab*, with its canonical forgetful functor, is a one-sorted algebraic category. In fact, it is concretely equivalent to the category of algebras for $(\mathcal{T}_{ab}, T_{ab})$: the functor

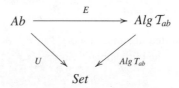

which to every group G assigns the algebra $\widehat{G}: \mathcal{T}_{ab} \to Set$ of Example 1.6, is concrete. And we have the concrete functor

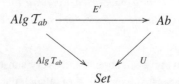

which to every algebra $A: \mathcal{T}_{ab} \to Set$ assigns the group G with $A \simeq \widehat{G}$ from 1.6. It is easy to verify that both $E \cdot E'$ and $E' \cdot E$ are naturally isomorphic to the identity functors.

11.16 Remark Given a concrete category $(\mathcal{A}, U)$, all subcategories of $\mathcal{A}$ are considered to be concrete using the domain restriction of U.

11.17 Proposition *Every variety of T-algebras for a one-sorted theory $(\mathcal{T}, T)$ is a one-sorted algebraic category.*

Proof Let $\sim$ be a congruence on $\mathcal{T}$ and let $Q: \mathcal{T} \to \mathcal{T}/\sim$ be the corresponding quotient functor. Since Q preserves finite products, $(\mathcal{T}/\sim, Q \cdot T)$ is a one-sorted theory. Consider the full subcategory $\mathcal{A}$ of $Alg\,\mathcal{T}$ specified by $\sim$. The equivalence

$$E: Alg\,(\mathcal{T}/\sim) \to \mathcal{A}$$

constructed in 10.15 is a concrete functor (because $Alg\,Q$ is concrete). Moreover, E is an isomorphism (not just an equivalence) because it is bijective on objects. Thus the inverse functor $E^{-1}: \mathcal{A} \to Alg\,(\mathcal{T}/\sim)$ is concrete. $\square$

11.18 Example The category *Graph* of graphs is algebraic but not one-sorted algebraic. In fact, a terminal object in *Graph* is the graph with one vertex and one edge, and it has a proper subobject given by the graph G with one vertex and no edge. Observe that G is neither terminal nor initial in *Graph*. Now use the following

11.19 Lemma *In a one-sorted algebraic category, a terminal object A has no nontrivial subobjects: for every subobject $m: B \to A$, B is either an initial object or a terminal one.*

Proof Given a one-sorted algebraic theory $(\mathcal{T}, T)$, denote by A a terminal object of $Alg\,\mathcal{T}$ and by I the initial object. Since $Alg\,T: Alg\,\mathcal{T} \to Set$ preserves limits (Corollary 11.9), $B(1)$ is a subobject of $A1 = 1$. If $B1 \simeq 1$, then m_1 is an isomorphism and thus m is an isomorphism (since $Alg\,T$ is conservative). If $B1 = \emptyset$, consider the unique monomorphism $a: I \to B$ and the induced map $a_1: I1 \to B1 = \emptyset$. Such a map is necessarily an isomorphism, and thus so is a. It is easy to see that concrete equivalences preserve the preceding property of terminal objects. $\square$

11.20 Example Even though the category of graphs is not a one-sorted algebraic category, the category *RGraph* of reflexive graphs is. Here the objects are directed graphs

$$G_e \overset{\tau}{\underset{\sigma}{\rightrightarrows}} G_v$$

together with a map $d\colon G_v \to G_e$ such that $\tau \cdot d = \mathrm{id}_{G_v} = \sigma \cdot d$. A morphism from $(G_e, G_v, \tau, \sigma, d)$ to $(G'_e, G'_v, \tau', \sigma', d')$ is a graph homomorphism,

$$(h_e\colon G_e \to G'_e, h_v\colon G_v \to G'_v),$$

such that $h_e \cdot d = d' \cdot h_v$. We consider this category as concrete over *Set* by $U(G_e, G_v, \tau, \sigma, d) = G_e$.

In fact, the category *RGraph* is concretely equivalent to the following one-sorted algebraic category $\mathcal{A}$: an object of $\mathcal{A}$ is a set G_e equipped with two maps,

$$G_e \overset{t}{\underset{s}{\rightrightarrows}} G_e$$

such that $s \cdot t = t$ and $t \cdot s = s$. A morphism from (G_e, t, s) to (G'_e, t', s') is a map $h_e\colon G_e \to G'_e$ such that $h_e \cdot t = t' \cdot h_e$ and $h_e \cdot s = s' \cdot h_e$.

1. Define $E\colon RGraph \to \mathcal{A}$ by assigning to a reflexive graph $(G_e, G_v, \tau, \sigma, d)$ the object (G_e, t, s) of $\mathcal{A}$ with $t = d \cdot \tau$ and $s = d \cdot \sigma$. This is a concrete functor.

2. Define $E'\colon \mathcal{A} \to RGraph$ by assigning to an object (G_e, t, s) in $\mathcal{A}$ the reflexive graph $(G_e, G_v, \tau, \sigma, d)$ by taking as $d\colon G_v \to G_e$ a joint equalizer of

$$G_e \overset{t}{\underset{s}{\overset{\mathrm{id}}{\rightrightarrows}}} G_e$$

This yields the canonical factorizations $\tau\colon G_e \to G_v$ of t through d and $\sigma\colon G_e \to G_v$ of s through d (such factorizations exist because $t = s \cdot t = t \cdot s \cdot t = t \cdot t$, and analogously for s). The rest of the definition of E' is straightforward. Again, E' is a concrete functor.

3. The verification that $E \cdot E'$ and $E' \cdot E$ are naturally isomorphic to the identity functors is easy.

11.21 Remark Let $(\mathcal{T}, T)$ be a one-sorted theory.

1. The forgetful functor $Alg\,T\colon Alg\,\mathcal{T} \to Set$ has a left adjoint. In fact, due to 4.11 being applied to $Y_\mathcal{N}\colon \mathcal{N}^{op} \to Set$, we can choose a left adjoint

$$F_T\colon Set \to Alg\,\mathcal{T}$$

in such a way that the square

$$\begin{array}{ccc} \mathcal{N}^{op} & \xrightarrow{T^{op}} & \mathcal{T}^{op} \\ {\scriptstyle Y_{\mathcal{N}}}\downarrow & & \downarrow{\scriptstyle Y_{\mathcal{T}}} \\ \text{Set} & \xrightarrow[F_T]{} & \text{Alg } \mathcal{T} \end{array}$$

commutes. Thus, for every natural number n,
  a. $F_T(n) = \mathcal{T}(n, -)$
  b. $F_T(\pi_i^n) = - \cdot T\pi_i^n$ for all $i = 0, \ldots, n-1$
2. The naturality square for η: Id $\to$ Alg $\mathcal{T} \cdot F_T$ applied to π_i^n yields the commutativity of

$$\begin{array}{ccc} 1 & \xrightarrow{\eta_1} & \mathcal{T}(1,1) \\ {\scriptstyle \pi_i^n}\downarrow & & \downarrow{\scriptstyle -\cdot T\pi_i^n} \\ n & \xrightarrow[\eta_n]{} & \mathcal{T}(n,1) \end{array}$$

that is,

$$\eta_n(i) = T\pi_i^n \quad \text{for all } i = 0, \ldots, n-1$$

(recall that π_i^n is the inclusion of i).
3. Since F_T preserves coproducts, $F_T X = \coprod_X Y_T(1)$ for every set X.
4. $\mathcal{T}$-algebras of the form $F_T X$, for X a set, are called *free algebras*. If X is finite, they are called *finitely generated free algebras*.

11.22 Corollary *Let* $(\mathcal{T}, T)$ *be a one-sorted theory.* $\mathcal{T}^{op}$ *is equivalent to the full subcategory of Alg* $\mathcal{T}$ *of finitely generated free algebras.*

Indeed, by the Yoneda lemma, $\mathcal{T}(n,k) \simeq \text{Alg } \mathcal{T}(Y_T(k), Y_T(n)) = \text{Alg } \mathcal{T}(F_T(k), F_T(n))$.

11.23 Remark 1. Since homomorphisms from $F_T 1$, the free $\mathcal{T}$-algebra on one generator, into an algebra form essentially the underlying set of that algebra, we obtain an algebraic theory $\overline{\mathcal{T}}$ whose morphisms in $\overline{\mathcal{T}}(k, 1)$ are precisely the elements of the free $\mathcal{T}$-algebra $F_T k$. Thus general hom-sets are given by

$$\overline{\mathcal{T}}(k, n) = (UF_T k)^n = UF_T k \times \cdots \times UF_T k$$

For example in case of abelian groups we have $F_T k = \mathbb{Z}^k$ and the elements of this free group can be represented by columns of k integers. The hom-set

$\overline{T}(k, n)$ consists of n-tuples of such columns, in other words, of $k \times n$ matrices of integers. Thus $\overline{T} = T_{ab}$, see 1.6.

In the concrete category of monoids we have the free monoid

$$F_T k = k^*$$

of all words over the set $k = \{0, 1, \ldots, k-1\}$. (The concatenation of words is the monoid operation, and the empty word is the neutral element.) Thus, the theory $\overline{T}$ has the hom-sets

$$\overline{T}(k, n) = k^* \times \cdots \times k^*$$

consisting of n-tuples of words over k.

2. Every object of a category with finite coproducts defines an algebraic theory $T(A)$ for which we need to fix coproduct injections:

$$p_i^n \colon A \to nA = A + \ldots + A \quad (n \text{ summands}).$$

The objects of $T(A)$ are natural numbers, and morphisms from n to k are the morphisms of $\mathcal{A}$ from $kA = A + \ldots + A$ to nA. Then $T \colon \mathcal{N} \to T(A)$ is defined by $T\pi_i^n = p_i^n$. Observe that $T(A)$ is equivalent to the full subcategory of $\mathcal{A}^{op}$ on all finite copowers of A under the equivalence functor $n \mapsto nA$. The corresponding category of $T(A)$-algebras can be equivalent to $\mathcal{A}$, as we have seen in the example $\mathcal{A} = Ab$ and $A = \mathbb{Z}$.

11.24 Remark Extending Remark 11.21, for every one-sorted algebraic category $(\mathcal{A}, U)$ with a left adjoint $F \dashv U$, a one-sorted theory can be constructed from the full subcategory of $\mathcal{A}^{op}$ on the objects $F\{x_0, \ldots, x_{n-1}\}$. Here a set of standard variables $x_0, x_1, x_2, \ldots$ is assumed. In fact, the n injections

$$\{x_0\} \to \{x_0, \ldots, x_{n-1}\}, \quad x_0 \mapsto x_i$$

define n morphisms $p_i^n \colon F\{x_0, \ldots, x_{n-1}\} \to F\{x_0\}$ in $\mathcal{A}^{op}$. Let T be the category whose objects are the natural numbers and whose morphisms are

$$T(n, k) = \mathcal{A}(F\{x_0, \ldots, x_{k-1}\}, F\{x_0, \ldots, x_{n-1}\}).$$

The composition in T is inherited from $\mathcal{A}^{op}$, and so are the identity morphisms. The functor $T \colon \mathcal{N} \to T$ is determined by the preceding choice of morphisms p_i^n for all $i \leq n$.

11.25 Example For the one-sorted theory (T_{ab}, T_{ab}) (Example 11.5), the induced adjunction $F_{T_{ab}} \dashv Alg\, T_{ab}$ is, up to concrete equivalence, the usual adjunction given by free abelian groups.

Using free algebras, we can restate some facts from Chapter 5:

11.26 Proposition *Let $(\mathcal{T}, T)$ be a one-sorted algebraic theory:*

1. *Free algebras are precisely the coproducts of representable algebras.*
2. *Every algebra is a regular quotient of a free algebra.*
3. *Regular projectives are precisely the retracts of free algebras.*

Proof
1. Following Remark 11.21, every free algebra is a coproduct of representable algebras. Conversely, every representable algebra is free by 11.21. Thus every coproduct of representable algebras is free because F_T preserves coproducts.

2. Following 4.2, every algebra is a regular quotient of a coproduct of representable algebras and then, by 1, a regular quotient of a free algebra.

3. This follows from point 1 and 5.14.2. □

11.27 Remark We have used the term *finitely generated* in two different situations: for objects of a category (see 5.21), and as a denotation of $F_T X$ with X finite. The following proposition demonstrates that there is no conflict. Note, however, that a finitely generated free algebra can, in principle, coincide with a nonfinitely generated one. In fact, in the variety of algebras satisfying, for a pair x, y of distinct variables, the equation $x = y$, all algebras are isomorphic.

11.28 Proposition *Let $(\mathcal{T}, T)$ be a one-sorted algebraic theory:*

1. *Finitely generated free algebras are precisely the representable algebras. These are precisely the free algebras that are finitely generated objects (in the sense of Definition 5.21).*
2. *Perfectly presentable algebras are precisely the retracts of finitely generated free algebras.*
3. *Finitely presentable algebras are precisely the coequalizers of (reflexive) pairs of morphisms between finitely generated free algebras.*

Proof
Point 1 follows from 11.21 and the observation that whenever the object $F_T X$ is finitely generated, it is isomorphic to $F_T X'$ for some finite set X'. In fact, the set X is the directed union of its nonempty finite subsets X'. Since F_T is a left adjoint and directed unions are directed colimits in *Set*, we see that $F_T X$ is a directed colimit with the colimit cocone formed by all Fi, where $i\colon X' \to X$ are the inclusion maps. Moreover, since each i is a split monomorphism in *Set*, $F_T X$ is the directed union of the finitely generated free algebras $F_T X'$. Since $F_T X$ is finitely generated, there exists a finite nonempty subset $i\colon X' \hookrightarrow X$ and a homomorphism $f\colon F_T X \to F_T X'$ such that $F_T i \cdot f = id_{F_T X}$. Thus

$F_T i \colon F_T X' \to F_T X$, being a monomorphism and a split epimorphism, is an isomorphism.

Point 2 follows from point 1 and 5.14.1, and point 3 follows from point 1 and 5.17. □

11.29 Remark Let $(\mathcal{T}, T)$ be a one-sorted algebraic theory and X a subset of the underlying set $A1$ of a $\mathcal{T}$-algebra A. The subalgebra of A generated by X (see Corollary 11.9) is a regular quotient of the free algebra $F_T X$. Indeed, consider the homomorphisms $\bar{i}_X \colon F_T X \to A$ corresponding to the inclusion $i_X \colon X \to A1$. Then $\bar{i}_X(F_T X)$ is a subalgebra of A because the forgetful functor preserves regular factorizations. This is obviously the least subalgebra of A containing X (use a diagonal fill-in; see 0.16), and the codomain restriction of $\bar{i}_X$ is a regular quotient.

11.30 Proposition *Let $(\mathcal{T}, T)$ be a one-sorted algebraic theory and A a $\mathcal{T}$-algebra. The following conditions are equivalent:*

1. *A is finitely generated (see 5.21).*
2. *A is a regular quotient of a finitely generated free algebra.*
3. *There exists a finite subset X of $A1$ not contained in any proper subalgebra of A.*

Proof The equivalence between points 1 and 2 follows from 5.21 and 11.18.1. For implication $1 \Rightarrow 3$ let A be a finitely generated object of $Alg\,\mathcal{T}$. Form a diagram in $Alg\,\mathcal{T}$ indexed by the poset of all finite subsets of $A1$ by assigning to every such $X \subseteq A1$ the subalgebra $\overline{X}$ of A generated by X (see 11.10). Given finite subsets X and Y with $X \subseteq Y \subseteq A1$, the connecting map $\overline{X} \to \overline{Y}$ is the inclusion map. Then the inclusion homomorphisms $i_X \colon \overline{X} \to A$ form a colimit cocone of this directed diagram. Since the functor $Alg\,\mathcal{T}(A, -)$ preserves this colimit, for $\mathrm{id}_A \in Alg\,\mathcal{T}(A, A)$ there exists a finite set X such that id_A lies in the image of i_X – but this proves $\overline{X} = A$.

For implication $3 \Rightarrow 2$ if $A = \overline{X}$ for a finite subset X of $A1$, then by 11.19, A is a regular quotient of the finitely generated free algebra $F_T X$. □

11.31 Remark

1. Recall the notion of an equivalence relation on an object A in a category from 3.12. If the category is $Alg\,\mathcal{T}$, we speak (as usual in general algebra) about *congruence on the algebra A* (instead of equivalence). This is a slight abuse of terminology since congruences were previously used for the theory $\mathcal{T}$ itself.

2. Similarly to Remark 11.10, for every T-algebra A and every subset X of $A1 \times A1$, there exists the least congruence on A whose underlying set contains X. Such a congruence is called the *congruence generated by X*.
3. The *finitely generated congruences* are those generated by finite subsets of $A1 \times A1$.

11.32 Lemma *Let $(\mathcal{T}, T)$ be a one-sorted algebraic theory. Given homomorphisms*

$$F_T X \xrightarrow[v]{u} B$$

in Alg $\mathcal{T}$, the congruence on B generated by the image of $\langle u, v \rangle$ coincides with that generated by the image of $\langle u_1 \cdot \eta_X, v_1 \cdot \eta_X \rangle$, where $\eta_X \colon X \to (F_T X)(1)$ is the unit of the adjunction $F_T \dashv Alg\, T$.

Proof Let R be the congruence generated by the image of $\langle u, v \rangle$ and S the congruence generated by the image of $\langle u_1 \cdot \eta_X, v_1 \cdot \eta_X \rangle$. To check the inclusion $R \subseteq S$, use the universal property of η_X and the diagonal fill-in (cf. 0.16). The other inclusion is obvious. □

11.33 Corollary *Let $(\mathcal{T}, T)$ be a one-sorted algebraic theory. A T-algebra A is finitely presentable iff there exists a coequalizer of the form*

$$R \xrightarrow[r_2]{r_1} F_T Y \xrightarrow{c} A$$

with Y a finite set and R a finitely generated congruence on $F_T Y$.

In fact, this follows from Proposition 11.28 and Lemma 11.32 by taking $B = F_T Y$ with X and Y finite sets.

As announced at the beginning of Chapter 10, we are going to prove that closure under directed unions in Birkhoff's variety theorem 10.22 can be avoided when the theory is one-sorted:

11.34 Proposition *Let $(\mathcal{T}, T)$ be a one-sorted algebraic theory. If a full subcategory $\mathcal{A}$ of Alg $\mathcal{T}$ is closed under products, subalgebras, and regular quotients, then it is a variety.*

Proof By 10.22, all we need is to prove that $\mathcal{A}$ is closed under directed unions. Let $A = \cup_{i \in I} A_i$ be a directed union of algebras in $\mathcal{A}$. Since the forgetful functor $Alg\, T$ preserves directed unions, $A1 = \cup_{i \in I}(A_i 1)$.

1. If $A1 = \emptyset$, then A is a subalgebra of any $\mathcal{T}$-algebra. Since $\mathcal{A}$ is nonempty (being closed in $Alg\,\mathcal{T}$ under products), this proves that A is in $\mathcal{A}$.

2. If $A1 \neq \emptyset$, we can choose $i_0 \in I$ such that $A_{i_0}1 \neq \emptyset$. The product $\prod_{i \geq i_0} A_i$ lies in $\mathcal{A}$. It assigns to $n \in \mathcal{N}$ the set of all tuples $(x_i)_{i \geq i_0}$ with $x_i \in A_i n$; we call the tuple *stable* if there exists $j \geq i_0$ such that $x_i = x_j$ for all $i \geq j$ (this makes sense because if $j \leq i$, then $A_j n \subseteq A_i n$), and we call x_j the *stabilizer* of the tuple. The stabilizer of a stable tuple is unique because I is directed. Define a subfunctor B of $\prod_{i \geq i_0} A_i$ by assigning to every n the set of all stable tuples in $\prod_{i \geq i_0} A_i n$. It is clear that since each A_i preserves finite products, so does B. Thus B is a subalgebra of the product, which proves that B lies in $\mathcal{A}$. We have a natural transformation $f \colon B \to A$ assigning to every stable tuple $(x_i)_{i \geq i_0}$ its stabilizer. Our choice of i_0 is such that f_1 is surjective: for a given $x \in A1$, choose $i_1 \in I$ such that $x \in A_{i_1}1$; since I is directed, there exists $j \in I$ with $i_0, i_1 \leq j$, and we can construct a stable tuple with stabilizer x. Following Corollary 11.9, f is a regular epimorphism. This proves that A lies in $\mathcal{A}$. □

Recall the algebraic duality of Chapter 9: if we restrict algebraic theories to the canonical ones, we obtain a contravariant biequivalence between the 2-category of algebraic categories and the 2-category of algebraic theories. In the one-sorted case, a better result is obtained since we do not have to restrict the theories at all.

11.35 Definition

1. The *2-category Th^1 of one-sorted theories* has
    objects: one-sorted algebraic theories
    1-cells: morphisms of one-sorted algebraic theories
    2-cells: natural transformations
2. The *2-category ALG^1 of one-sorted algebraic categories* has
    objects: one-sorted algebraic categories
    1-cells: concrete functors
    2-cells: natural transformations

11.36 Remark Every 1-cell in ALG^1 is a faithful and conservative algebraic functor.

11.37 Definition We denote by

$$Alg^1 \colon (Th^1)^{op} \to ALG^1$$

the 2-functor assigning to every one-sorted theory $(\mathcal{T}, T)$ the concrete category $Alg^1(\mathcal{T}, T) = (Alg\,T \colon Alg\,\mathcal{T} \to Set)$, to every 1-cell $M \colon (\mathcal{T}_1, T_1) \to (\mathcal{T}_2, T_2)$ the concrete functor $Alg^1 M = (-) \cdot M$, and to every 2-cell $\alpha \colon M \to N$ the natural transformation $Alg^1\alpha: Alg^1M \to Alg^1N$ whose component at a $\mathcal{T}_2$-algebra A is $A \cdot \alpha: A \cdot M \to A \cdot N$.

11.38 Theorem: *One-sorted algebraic duality* The 2-category ALG^1 of one-sorted algebraic categories is biequivalent to the dual of the 2-category Th^1 of one-sorted algebraic theories. In fact, the 2-functor

$$Alg^1: (Th^1)^{op} \to ALG^1$$

is a biequivalence.

Proof

1. Alg^1 is well defined and essentially surjective (in the sense of the 2-category ALG^1, which means surjectivity up to concrete equivalence) by definition of one-sorted algebraic category.

2. We will prove that for two one-sorted algebraic theories $(\mathcal{T}_1, T_1)$ and $(\mathcal{T}_2, T_2)$, the functor

$$Th^1((\mathcal{T}_1, T_1), (\mathcal{T}_2, T_2)) \xrightarrow{Alg^1_{(\mathcal{T}_1,T_1),(\mathcal{T}_2,T_2)}} ALG^1((Alg\,\mathcal{T}_2, Alg\,T_2), (Alg\,\mathcal{T}_1, Alg\,T_1))$$

is an equivalence of categories. The proof that $Alg^1_{(\mathcal{T}_1,T_1),(\mathcal{T}_2,T_2)}$ is full and faithful is the same as in Theorem 9.15. It remains to be proven that $Alg^1_{(\mathcal{T}_1,T_1),(\mathcal{T}_2,T_2)}$ is essentially surjective: consider a concrete functor

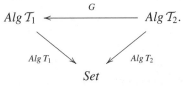

It is our task to find a theory morphism $M: (\mathcal{T}_1, T_1) \to (\mathcal{T}_2, T_2)$ with $G \simeq Alg^1 M$. We have the left adjoint F_T of Remark 11.21, and we denote by $F: Alg\,\mathcal{T}_1 \to Alg\,\mathcal{T}_2$ a left adjoint of G. The commutativity of the preceding triangle yields a natural isomorphism

$$\psi: F_{T_2} \to F \cdot F_{T_1}.$$

We are going to prove that $F \cdot Y_{\mathcal{T}_1}$ factorizes (up to natural isomorphism) through $Y_{\mathcal{T}_2}$:

$$\begin{array}{ccc} \mathcal{T}_1^{op} & \xrightarrow{Y_{\mathcal{T}_1}} & Alg\,\mathcal{T}_1 \\ {\scriptstyle M^{op}}\downarrow & \simeq & \downarrow{\scriptstyle F} \\ \mathcal{T}_2^{op} & \xrightarrow{Y_{\mathcal{T}_2}} & Alg\,\mathcal{T}_2 \end{array} \qquad (11.1)$$

We define $M: \mathcal{T}_1 \to \mathcal{T}_2$ to be the identity on objects, and for a morphism $f: t \to s$ in $\mathcal{T}_1$ we define Mf as follows. Since $Y_{\mathcal{T}_2}$ is full and faithful, there exists a unique morphism $Mf: Mt \to Ms$ such that the following diagram commutes:

$$Y_{\mathcal{T}_2}M(s) = Y_{\mathcal{T}_2}T_2(s) = F_{T_2}(s) \xrightarrow{\psi_s} FF_{T_1}(s) = FY_{\mathcal{T}_1}T_1(s) = FY_{\mathcal{T}_1}(s)$$

$$\downarrow{Y_{\mathcal{T}_2}M(f)} \qquad\qquad\qquad\qquad\qquad\qquad \downarrow{FY_{\mathcal{T}_1}(f)}$$

$$Y_{\mathcal{T}_2}M(t) = Y_{\mathcal{T}_2}T_2(t) = F_{T_2}(t) \xrightarrow{\psi_t} FF_{T_1}(t) = FY_{\mathcal{T}_1}T_1(t) = FY_{\mathcal{T}_1}(t)$$

$$(11.2)$$

The equalities $F_{T_i} \cdot Y_{\mathcal{N}} = Y_{\mathcal{T}_i} \cdot T_i^{op}$ ($i = 1, 2$) for the embedding $Y_{\mathcal{N}}: \mathcal{N}^{op} \to Set$ come from Remark 11.21. The functoriality of

$$M: \mathcal{T}_1 \to \mathcal{T}_2$$

follows from the uniqueness of Mf. Moreover, since the isomorphism ψ is natural, if $f = T_1 g$ for some $g: t \to s$ in $\mathcal{N}$, then $Mf = T_2 g$. Diagram (11.2) gives also the natural isomorphism needed in Diagram (11.1). This finishes the proof: $F \simeq M^*$ by 9.3, and then $G \simeq Alg\, M$. □

11.39 Remark One can modify this duality by restricting to uniquely transportable algebraic categories (13.16). One gets a dual equivalence (rather than a biequivalence). This new duality excludes the categories $Alg\, \mathcal{T}$ as such but replaces them with equivalent categories consisting of algebras for equational theories. This result, established in Appendix C, gives an alternative approach to the classical duality between finitary monads and finitary monadic categories over *Set* presented in Appendix A.

Historical remarks

In his dissertation Lawvere (1963) introduced the name *algebraic category* for one equivalent to the category $Alg\, \mathcal{T}$ of algebras of a one-sorted algebraic theory. Our decision to use *concrete* equivalence is motivated by the precise analogy one gets to finitary monadic categories over *Set* (see Appendix A).

Another variant, based on pseudoconcrete functors in place of concrete ones, is to take all categories pseudoconcretely equivalent to the categories $Alg\, \mathcal{T}$ given earlier. This is briefly mentioned in Appendix C.

12
Algebras for an endofunctor

The aim of this chapter is to show how limits and colimits are constructed in categories of algebras for a finitary endofunctor of *Set* and, in particular, in the category of Σ-algebras. These results are actually true for finitary endofunctors of all locally finitely presentable categories, and the proofs are the same. In the special case of *Set*, we also characterize the finitary endofunctors as precisely the quotients of the polynomial endofunctors H_Σ. We will prove in 13.23 that the category of algebras for a finitary endofunctor of *Set* is a one-sorted algebraic category.

12.1 Remark

1. The concept of H-algebra in 2.25 can be formulated for the endofunctor H of an arbitrary category $\mathcal{K}$: it is a pair (A, a) consisting of an object A and a morphism $a\colon HA \to A$. The category

$$H\text{-}Alg$$

has as objects H-algebras and as morphisms from (A, a) to (B, b) those morphisms $f\colon A \to B$ for which $f \cdot a = b \cdot Hf$.
2. We denote by

$$U_H\colon H\text{-}Alg \to \mathcal{K}$$

the canonical forgetful functor $(A, a) \mapsto A$.
3. In this chapter, we restrict ourselves to finitary endofunctors H of *Set*. See also Remark 12.17.

12.2 Remark
We want to show how colimits of H-algebras are obtained. We begin with the simplest case: the initial H-algebra. We will prove that it can be

obtained by iterating the unique morphism $u\colon \emptyset \to H\emptyset$. More precisely, let us form the ω-chain

$$\emptyset \xrightarrow{u} H\emptyset \xrightarrow{Hu} H^2\emptyset \xrightarrow{H^2u} H^3\emptyset \longrightarrow \cdots$$

We call it the *initial chain* of H. Its colimit

$$I = \colim_{n\in\mathbb{N}} H^n\emptyset$$

carries the structure of an H-algebra. Indeed, since H preserve colimits of ω-chains,

$$HI \simeq \colim_{n\in\mathbb{N}} H(H^n\emptyset) \simeq \colim_{n\in\mathbb{N}} H^n\emptyset = I.$$

We denote by $i\colon HI \to I$ the canonical isomorphism. In more detail, denote by

$$v_n\colon H^n\emptyset \to I \quad (n \in \mathbb{N})$$

a colimit cocone for I. Then $i\colon HI \to I$ is defined by

$$i \cdot Hv_{n-1} = v_n \quad \text{for all } n \geq 1.$$

12.3 Lemma *The H-algebra $i\colon HI \to I$ is initial.*

Proof For every algebra $a\colon HA \to A$, define a cocone $f_n\colon H^n\emptyset \to A$ of the initial chain as follows: $f_0\colon \emptyset \to A$ is unique and

$$f_{n+1} = a \cdot Hf_n \colon HH^n\emptyset \to A.$$

The unique morphism $f\colon I \to A$ with $f \cdot v_n = f_n$ ($n \in \mathbb{N}$) is a homomorphism: since the cocone (Hv_n) is a colimit cocone, and thus collectively epimorphic, this follows from the commutative diagram

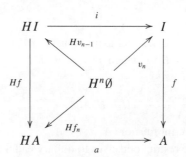

Conversely, if $f\colon I \to A$ is a homomorphism, the preceding diagram proves that for every $n \geq 1$, the previous morphism f_n is equal to $f \cdot v_n$. This shows the uniqueness. □

12.4 Example We describe the initial H_Σ-algebra for the polynomial functor

$$H_\Sigma X = \coprod_{k \in \mathbb{N}} \Sigma_k \times X^k$$

by applying labeled trees.

Recall that *a tree* is a directed graph with a distinguished node (root) such that for every node, there exists a unique path from the root into it. We work with *ordered trees*; that is, for every node, the set of children nodes is linearly ordered (from left to right). Trees are identified when there exist an isomorphism respecting the ordering and the labels.

For every signature Σ, by a Σ-*tree* is meant a tree labeled in Σ so that every node has the number of children equal to the arity of its label.

Concerning *initial H_Σ-algebra*; we can represent $H_\Sigma \emptyset = \Sigma_0$ by the set of all singleton trees labeled by elements of Σ_0. Given a tree representation of $H_\Sigma^k \emptyset$, we represent

$$H_\Sigma^{k+1} \emptyset = \coprod_{n \in \mathbb{N}} \Sigma_n \times (H_\Sigma^k \emptyset)^n$$

by the set of all trees

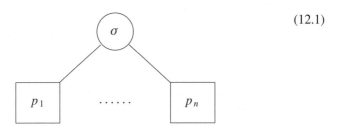

(12.1)

with $\sigma \in \Sigma_n$ and $p_1, \ldots, p_n \in H_\Sigma^k \emptyset$. In this way, we see that for every $k \in \mathbb{N}$,

$$H_\Sigma^k \emptyset = \text{all } \Sigma\text{-trees of depths less than } k.$$

The preceding ω-chain is the chain of inclusion maps $\emptyset \subseteq H_\Sigma \emptyset \subseteq H_\Sigma^2 \emptyset \subseteq \ldots$, and its colimit

$$I = \bigcup_{k \in \mathbb{N}} H_\Sigma^k(\emptyset)$$

is the set of all finite Σ-trees. The algebraic structure $i \colon H_\Sigma I \to I$ is given by tree tupling: to every n-tuple of trees corresponding to the summand of $\sigma \in \Sigma_n$, it assigns the tree (12.1).

12.5 Remark. Free H-algebras Let H be a finitary endofunctor of *Set*. We now describe free H-algebras, that is, a left adjoint of the forgetful functor U_H. For every set X, the endofunctor

$$H(-) + X$$

is also finitary. Therefore, following Lemma 12.3, it has an initial algebra.

12.6 Proposition *The free H-algebra on X is the initial algebra for the endofunctor $H(-) + X$.*

Explicitly, if H^*X is the initial algebra for $H(-) + X$ with structure

$$i_X \colon HH^*X + X \to H^*X,$$

then the components

$$\varphi_X \colon HH^*X \to H^*X \quad \text{and} \quad \eta_X \colon X \to H^*X$$

of i_X form the algebra structure and the universal morphism, respectively.

Proof This follows easily from the observation that to specify an algebra for $H(-) + X$ on a set A means to specify an algebra $HA \to A$ for H and a function $X \to A$. □

12.7 Corollary *The free H-algebra on X is the colimit of the ω-chain*

$$\emptyset \to H\emptyset + X \to H(H\emptyset + X) + X \to \ldots.$$

12.8 Notation $F_H \colon Set \to H\text{-}Alg$ denotes the left adjoint of U_H. In case $H = H_\Sigma$, we use F_Σ instead of F_{H_Σ}.

12.9 Example We describe the free H_Σ-algebra on a set X. Observe that

$$H_\Sigma(-) + X = H_{\overline{\Sigma}}$$

for the signature $\overline{\Sigma}$ obtained from Σ by adding nullary operation symbols from X. Thus the description of initial algebra in 12.4 immediately yields a description of the free H_Σ-algebra $F_\Sigma X$ on X as the algebra of all finite Σ-*trees on X*, that is, finite labeled trees with leaves labeled in $X + \Sigma_0$, and nodes with $n > 0$ children labeled in Σ_n. The operations of $F_\Sigma X$ are given by tree tupling. The universal morphism assigns to a variable $x \in X$ the singleton tree labeled by x.

12.10 Example For the signature Σ of a single binary operation $*$, we have a description of $F_\Sigma X$ for $X = \{p_1, p_2, p_3\}$ as all binary trees with leaves labeled by p_1, p_2, p_3. Examples follow:

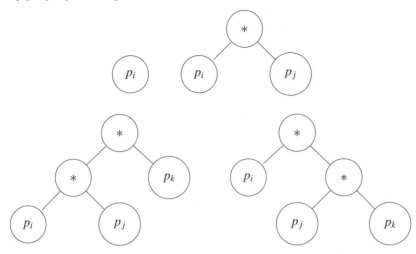

12.11 Example A commutative binary operation can be expressed by the functor H assigning to every set X the set HX of all unordered pairs in X and to every function f the function Hf acting as f component-wise.

An H-algebra is a set with a commutative binary operation. The free algebra on X is the colimit of the chain

$$\emptyset \to H\emptyset + X = X \to HX + X \to H(HX + X) + X \to \ldots.$$

We can represent the elements of HZ as binary nonordered trees with both subtree elements of Z, and we see that the nth set in the preceding chain consists of precisely all binary nonordered trees of depth less than n with leaves labeled in X. Consequently, the initial algebra is

$$IX = \text{all unordered binary trees over } X.$$

12.12 Proposition *For every finitary endofunctor H on Set, the category H-Alg has limits and sifted colimits preserved by the forgetful functor.*

We know from 6.30 that H preserves sifted colimits. We can generalize the statement about sifted colimits to say: for every type of colimits preserved by H, the category H-Alg has colimits of that type preserved by U_H.

Proof We prove the more general formulation about colimits; the proof of limits is analogous. Let $D: \mathcal{D} \to H\text{-}Alg$ be a diagram with objects $Dd = (A_d, a_d)$, and let $A = \text{colim } A_d$ be the colimit of $U_H \cdot D$ in Set with the colimit cocone $c_d: A_d \to A$. If H preserves this colimit, there exists a unique

H-algebra structure $a\colon HA \to A$ turning each c_d into a homomorphism. In fact, the commutative squares

$$\begin{array}{ccc} HA_d & \xrightarrow{Hc_d} & HA \\ a_d \downarrow & & \downarrow a \\ A_d & \xrightarrow{c_d} & A \end{array}$$

define a unique a since (1) $c_d \cdot a_d$ is a cocone of $U_H \cdot D$ and (2) Hc_d is the colimit cocone of $H \cdot U_H \cdot D$. It is easy to see that the algebra (A, a) is a colimit of D in H-Alg with the cocone c_d. $\square$

12.13 Theorem *For every finitary endofunctor H on Set, the category H-Alg is cocomplete. It also has regular factorizations of morphisms preserved by the forgetful functor.*

Proof
1. We start with the latter statement. Observe that regular epimorphisms split in *Set*, and therefore H preserves them. Given a homomorphism $h\colon (A, a) \to (B, b)$ in H-Alg and a factorization $h = m \cdot e$ with

$e\colon A \to C$ a regular epimorphism and $m\colon C \to B$ a monomorphism

in *Set*, use the diagonal fill-in to obtain an H-algebra structure $c\colon HC \to C$, turning e and m into homomorphisms:

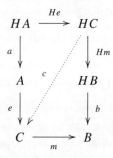

Since U_H is faithful, m is a monomorphism in H-Alg, and e is a regular epimorphism in H-Alg because given a pair $u, v\colon X \rightrightarrows A$ with coequalizer e in *Set*, the corresponding homomorphisms $\overline{u}, \overline{v}\colon F_H X \to A$ from the free H-algebra have the coequalizer e in H-Alg.

2. Arguing as in 4.1, to prove that H-Alg is cocomplete, it is sufficient to prove that it has finite coproducts. Since by 12.3, H-Alg has an initial object, it remains to consider binary coproducts. Thus we are to prove that the diagonal functor $\Delta\colon H\text{-}Alg \to H\text{-}Alg \times H\text{-}Alg$ has a left adjoint (see 0.11). Since by Proposition 12.12, the category H-Alg is complete, it is sufficient (using the adjoint functor theorem 0.8) to find a solution set for every pair (A_1, a_1), (A_2, a_2) of H-algebras; that is, we need a set of cospans

$$(A_1, a_1) \xrightarrow{f_1} (C, c) \xleftarrow{f_2} (A_2, a_2)$$

in H-Alg through which all cospans factorize. Consider the coproduct

$$A_1 \xrightarrow{c_1} A_1 + A_2 \xleftarrow{c_2} A_2$$

in Set, denote by $f\colon A_1 + A_2 \to C$ the morphism induced by the cospan (f_1, f_2), and let $\overline{f}\colon F_H(A_1 + A_2) \to (C, c)$ be the homomorphism corresponding to f by adjunction. We claim that a solution set is provided by those cospans (f_1, f_2) such that $\overline{f}$ is a regular quotient of $F_H(A_1 + A_2)$. This is indeed a set because $\overline{f}$ is a regular epimorphism also in Set. For any cospan

$$(A_1, a_1) \xrightarrow{g_1} (D, d) \xleftarrow{g_2} (A_2, a_2),$$

consider the regular factorization in H-Alg:

$$F_H(A_1 + A_2) \xrightarrow{\overline{g}} (D, d)$$
$$\searrow e \qquad \nearrow m$$
$$(C, c)$$

We get a new cospan in Set by defining

$$f_i\colon A_i \xrightarrow{c_i} A_1 + A_2 \xrightarrow{\eta_{A_1+A_2}} F_H(A_1 + A_2) \xrightarrow{e} C \quad (i = 1, 2)$$

(where η is the unit of the adjunction $F_H \dashv U_H$). The cospan (g_1, g_2) factorizes through (f_1, f_2) because $g_i = m \cdot f_i$. Moreover, (f_1, f_2) is a cospan in H-Alg: this follows easily from the fact that g_i and m are homomorphisms of H-algebras and m is a monomorphism in Set. Finally, (f_1, f_2) has the desired property because $\overline{f} = e$. □

12.14 Remark So far, we have mentioned, besides the polynomial functors H_Σ, only one finitary functor that is not polynomial (see Example 12.11), and that functor is an obvious quotient of the polynomial functor $H_\Sigma X = X \times X$.

In general, a *quotient* of a functor H is represented by a natural transformation $\alpha: H \to \overline{H}$ with epimorphic components. We now prove that finitary endofunctors of *Set* are indeed precisely the quotients of the polynomial ones.

12.15 Theorem *For an endofunctor H on Set, the following conditions are equivalent:*

1. *H is finitary.*
2. *H is a quotient of a polynomial functor.*
3. *Every element of HX lies in the image of Hi for the inclusion $i: Y \to X$ of a finite subset Y.*

Proof For $3 \Rightarrow 1$; Let

$$D: \mathcal{D} \to Set$$

be a filtered diagram with a colimit cocone

$$c_d: Dd \to C \quad (d \in \operatorname{obj} \mathcal{D}).$$

We prove that the diagram $D \cdot H$ has the colimit

$$Hc_d: HDd \to HC$$

in *Set*. For that, by 0.6, it is sufficient to verify that

a. every element x of HC lies in the image of Hc_d for some d
b. given elements $y_1, y_2 \in HDd$ merged by Hc_d, there exists a connecting morphism $f: d \to d'$ of $\mathcal{D}$ with Hf also merging y_1 and y_2

For point a, choose a finite subset $i: Y \to X$ with x lying in the image of Hi. Since $C = \operatorname{colim} Dd$ is a filtered colimit in *Set*, there exists a factorization

$$i = c_d \cdot j \quad \text{for some } d \in \operatorname{obj} \mathcal{D} \text{ and } j: Y \to Dd.$$

Thus x lies in the image of Hc_d.

For point b, choose a finite subset $i: Z \to Dd$ such that y_1, y_2 lie in the image of Hi. Since $C = \operatorname{colim} Dd$ is a filtered colimit, there exists a connecting morphism $f: d \to d'$ such that the domain restriction of $c_{d'}$ to the image

$i'\colon Z' \to Dd'$ of $f \cdot i$ is a monomorphism. We obtain a commutative diagram

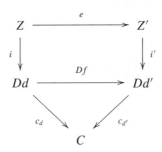

for some morphism e such that $c_{d'} \cdot i'$ is a monomorphism. Without loss of generality, $Z' \neq \emptyset$ provided $C \neq \emptyset$. Then $c_{d'} \cdot i'$ is a split monomorphism. Then $H(c_{d'} \cdot i')$ also is a split monomorphism, and we conclude that HDf merges y_1, y_2.

Implication 1 $\Rightarrow$ 3 is obvious from the description of filtered colimits in *Set* (see 0.6).

Implication 1 $\Rightarrow$ 2 follows from the Yoneda lemma. Define the signature Σ by using, for every $n \in \mathbb{N}$, the elements of Hn as the operation symbols σ of arity $n \in \mathbb{N}$. In short, $Hn = \Sigma_n$. Then we have a natural transformation $\alpha\colon H_\Sigma \to H$ which, given an operation symbol $\sigma \in \Sigma_n$ (i.e., $\sigma \in Hn$), assigns to the corresponding n-tuple $f\colon n \to Z$ the value

$$\alpha_Z(\sigma(f)) = Hf(\sigma).$$

In other words, the component of α_Z restricted to the functor $Set(n, -)$ corresponding to $\sigma \in \Sigma_n$ is the Yoneda transformation of σ. Condition 3 tells us precisely that α_Z is a surjective map for every set Z.

For implication 2 $\Rightarrow$ 3; every polynomial functor satisfies condition 3. Indeed, to choose an element $x \in H_\Sigma X$ means to fix a symbol $\sigma \in \Sigma_n$ so that $x = (x_1, \ldots, x_n) \in X \times \ldots \times X$. Therefore, as Y, we can take $\{x_1, \ldots, x_n\}$.

Let now $\alpha\colon H_\Sigma \to H$ be a quotient, and fix an element $x \in HX$. Since $\alpha_X\colon H_\Sigma X \to HX$ is surjective, there exists $y \in H_\Sigma$ such that $\alpha_X(y) = x$. Find a finite subset $i\colon Y \to X$ such that $y = (H_\Sigma i)(\overline{y})$ for some $\overline{y} \in H_\Sigma$. By naturality of α, we have

$$x = \alpha_X(H_\Sigma i)(\overline{y}) = (Hi)\alpha_Y(\overline{y}).$$

$\square$

12.16 Remark In 13.23, we will see that for every presentation of a functor H as a quotient functor of H_Σ, the category of H-algebras can be viewed as an equational category of Σ-algebras.

12.17 Remark Most of the results in this chapter have an obvious generalization to endofunctors H of cocomplete categories $\mathcal{K}$ that preserve sifted colimits.

1. The initial chain of Remark 12.2 is defined by denoting by $\emptyset$ an initial object of $\mathcal{K}$ and using the unique morphism $u\colon \emptyset \to H\emptyset$. The corresponding H-algebra is initial.
2. The free H-algebra on an object X of $\mathcal{K}$ is the intial algebra for the endofunctor $H(-) + X$.
3. The category H-Alg is complete and cocomplete, and the forgetful functor into $\mathcal{K}$ preserves limits and sifted colimits.

Historical remarks

Algebras for an endofunctor were introduced by Lambek (1968). The initial algebra construction of Remark 12.2 and its free-algebra variation of Corollary 12.7 stem from Adámek (1974). Factorizations and colimits in categories H-Alg were studied by Adámek (1977). The fact that finitary endofunctors on Set yield one-sorted algebraic categories follows from the work of Barr (1970) on free monads (see Appendix A).

13
Equational categories of Σ-algebras

This chapter shows the precise relationship of the classical one-sorted general algebra and algebraic theories: we prove that every equational category of Σ-algebras is a one-sorted algebraic category (in the sense of Definition 11.13), and conversely, every one-sorted algebraic category can be presented by equations of Σ-algebras for some (one-sorted) signature. (The case of S-sorted signatures is treated in Chapter 14.)

We have introduced varieties in algebraic categories in Chapter 10. The classical equational categories, that is, full subcategories of Σ-algebras presented by equations, are a special case. In fact, we demonstrate that equations in the sense of pairs of terms over Σ canonically correspond to equations in the sense of Definition 10.1. We also prove that the categorical concepts of finitely presentable or finitely generated objects have in categories of Σ-algebras their classical meaning.

13.1 Remark We described a left adjoint

$$F_\Sigma: Set \to \Sigma\text{-}Alg$$

of the forgetful functor $U_\Sigma: \Sigma\text{-}Alg \to Set$ in 12.9 (recall from 2.25 that $H_\Sigma\text{-}Alg = \Sigma\text{-}Alg$). The more standard description is that $F_\Sigma X$ is the following Σ-*term-algebra*: the underlying set is the smallest set such that every element $x \in X$ is a Σ-term and for every $\sigma \in \Sigma$ of arity n and for every n-tuple of Σ-terms $p_1, \ldots, p_n$, we have a Σ-term $\sigma(p_1, \ldots, p_n)$. The Σ-algebra structure on $F_\Sigma X$ is given by the formation of terms $\sigma(p_1, \ldots, p_n)$. This defines a functor $F_\Sigma: Set \to \Sigma\text{-}Alg$ on objects. To define it on morphisms $f: X \to Z$, let $F_\Sigma f$ be the function which in every term p of

$F_\Sigma X$ substitutes for every variable $x \in X$ the variable $f(x)$. More explicitly; if $x \in X$, then $F_\Sigma f(x) = f(x)$; if $p_1, \ldots, p_n \in F_\Sigma X$ and $\sigma \in \Sigma_n$, then $F_\Sigma f(\sigma(p_1, \ldots, p_n)) = \sigma(F_\Sigma f(p_1), \ldots, F_\Sigma f(p_n))$. It is easy to verify that F_Σ is a well-defined functor that is naturally isomorphic to the Σ-tree functor of 12.9. Thus we have $F_\Sigma \dashv U_\Sigma$. The unit of the adjunction is the inclusion of variables into the set of Σ-terms: $\eta_X \colon X \to F_\Sigma X$.

13.2 Notation Suppose that a set of standard variables $x_0, x_1, x_2, \ldots$ is given. Then the free Σ-algebras

$$F_\Sigma\{x_0, \ldots, x_{n-1}\}$$

yield, by 11.24, a one-sorted theory for Σ-Alg. We denote this theory by

$$(\mathcal{T}_\Sigma, T_\Sigma).$$

Thus morphisms from n to 1 in $\mathcal{T}_\Sigma$ are the Σ-terms in variables $x_0, \ldots, x_{n-1}$. General hom-sets are given by k-tuples of these terms:

$$\mathcal{T}_\Sigma(n, k) = (F_\Sigma n)^k,$$

and $T_\Sigma \colon \mathcal{N} \to \mathcal{T}_\Sigma$ assigns to every function $g \colon k \to n$ in Set, that is, $g \in \mathcal{N}(n, k)$, the k-tuple of terms

$$x_{g(0)}, \ldots, x_{g(k-1)}.$$

13.3 Lemma *The category Σ-Alg is concretely equivalent to the category of algebras of $(\mathcal{T}_\Sigma, T_\Sigma)$.*

Proof Define $E \colon \Sigma$-$Alg \to Alg\,\mathcal{T}_\Sigma$ on objects as follows. For a Σ-algebra (A, a), the corresponding functor from $\mathcal{T}_\Sigma$ to Set is given on objects by $n \mapsto A^n$ and on morphisms $t \colon n \to 1$ by the function $A^n \to A$ of evaluation of the term t in the given algebra. This function takes a map $f \colon n \to A$ to $\overline{f}(t) \in A$, where $\overline{f} \colon F_\Sigma\{x_0, \ldots, x_{n-1}\} \to (A, a)$ is the unique homomorphism extending f.

Conversely, if B is a $\mathcal{T}_\Sigma$-algebra, we get a Σ-algebra structure on the set $B1$ as follows: if $\sigma \in \Sigma_n$, then $\sigma(x_0, \ldots, x_{n-1}) \in F_\Sigma n = \mathcal{T}_\Sigma(n, 1)$, and this yields an n-ary operation on $B1$ by applying B to that morphism (recall that Bn is isomorphic to the nth power of $B1$). This gives a concrete equivalence $E \colon \Sigma$-$Alg \to Alg\,\mathcal{T}_\Sigma$. □

13.4 Definition Given signatures Σ and Σ', a *morphism of signatures* is a function $f \colon \Sigma \to \Sigma'$ preserving the arities. This leads to the category of signatures $Sign$ – this is just the slice category $Set \downarrow \mathbb{N}$.

13.5 Definition For every one-sorted algebraic theory $(\mathcal{T}, T)$, we define the signature

$$\mathcal{C}(\mathcal{T}, T)$$

whose n-ary symbols are precisely the morphisms from n to 1 in $\mathcal{T}$:

$$(\mathcal{C}(\mathcal{T}, T))_n = \mathcal{T}(n, 1).$$

This construction can be easily extended to morphisms of one-sorted theories.

13.6 Example The signature $\mathcal{C}(\mathcal{T}_\Sigma, T_\Sigma)$ has all Σ-terms in variables $x_0, \ldots, x_{n-1}$ as n-ary operation symbols. Therefore there is a canonical morphism of signatures

$$\eta_\Sigma \colon \Sigma \longrightarrow \mathcal{C}(\mathcal{T}_\Sigma, T_\Sigma)$$

given by $\eta_\Sigma(\sigma) = \sigma(x_0, \ldots, x_{n-1}) \in F_\Sigma n$ for any σ of arity n.

13.7 Proposition: *A free one-sorted theory on a signature* For every signature Σ, the theory $(\mathcal{T}_\Sigma, T_\Sigma)$ is free on Σ; that is, given a one-sorted theory $(\mathcal{T}, T)$, for every morphism $G \colon \Sigma \to \mathcal{C}(\mathcal{T}, T)$ of signatures there exists a unique morphism $\overline{G} \colon (\mathcal{T}_\Sigma, T_\Sigma) \to (\mathcal{T}, T)$ of one-sorted theories such that $\mathcal{C}(\overline{G}) \cdot \eta_\Sigma = G$:

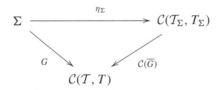

Proof

1. We define a functor $\overline{G} \colon \mathcal{T}_\Sigma \to \mathcal{T}$ on objects by $n \mapsto n$ and on morphisms $p \in \mathcal{T}_\Sigma(k, 1)$, that is, Σ-terms on $\{x_0, \ldots, x_{k-1}\}$, by structural induction:

   i. For variables $x_i \in \mathcal{T}_\Sigma(k, 1)$, put $\overline{G}x_i = T\pi_i^k$, the chosen projection in $\mathcal{T}(k, 1)$.

   ii. Given $p = \sigma(p_1, \ldots, p_n)$ where $\sigma \in \Sigma_n$, $p_i \in \mathcal{T}_\Sigma(k, 1)$ $(i = 1, \ldots, n)$ and $\overline{G}p_i$ are defined already, put

$$\overline{G}p \colon k \xrightarrow{\langle \overline{G}p_1, \ldots, \overline{G}p_n \rangle} n \xrightarrow{G\sigma} 1.$$

It is clear that $\overline{G} \cdot T_\Sigma = T$ and $\mathcal{C}(\overline{G}) \cdot \eta_\Sigma = G$.

2. Uniqueness: Let $M: (\mathcal{T}_\Sigma, T_\Sigma) \to (\mathcal{T}, T)$ be a morphism of one-sorted theories with $C(M) \cdot \eta_\Sigma = G$. Since M preserves finite products, all we have to prove is that it is determined when precomposed by η_Σ. This is the case, indeed:

i. For variables $x_i \in \mathcal{T}_\Sigma(k, 1)$, use $M \cdot T_\Sigma = T$ so that $M(\eta_\Sigma(x_i))$ is the ith projection π_i^k.

ii. Consider $p = \sigma(p_1, \ldots, p_n)$ with $\sigma \in \Sigma_n$ and $p_i \in \mathcal{T}_\Sigma(k, 1)$ ($i = 1, \ldots, n$). Since $\sigma(p_1, \ldots, p_n) = \sigma(x_1, \ldots, x_n) \cdot \langle p_1, \ldots, p_n \rangle$ in $\mathcal{T}_\Sigma$, we have that $Mp = M(\eta_\Sigma(\sigma)) \cdot \langle Mp_1, \ldots, Mp_m \rangle$. □

13.8 Remark Let $(\mathcal{T}, T)$ be a one-sorted theory. If we apply the construction described in the first part of the proof of 13.7 to the identity morphism on $C(\mathcal{T}, T)$, we get a morphism

$$\varepsilon_{(\mathcal{T},T)}: \mathcal{T}_{C(\mathcal{T},T)} \longrightarrow (\mathcal{T}, T)$$

of one-sorted theories. It is clearly full, and then, by 10.13, the unique functor ε' making commutative the following diagram of morphisms of one-sorted theories is an isomorphism:

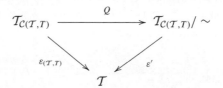

Therefore

1. $(\mathcal{T}, T)$ is a quotient of the free one-sorted theory $\mathcal{T}_{C(\mathcal{T},T)}$.
2. $Alg\,\mathcal{T}$ is a variety of $C(\mathcal{T}, T)$-algebras.

To improve the previous result, we need the notion of equational category of Σ-algebras. We start comparing the classical notion of equation with the one introduced in Chapter 10.

13.9 Remark

1. Classically, equations are expressions

$$t = t',$$

where t and t' are terms in variables $x_0, \ldots, x_{n-1}$ for some n. This is a special case of 10.1: here we have a parallel pair $t, t': n \rightrightarrows 1$ in the theory $\mathcal{T}_\Sigma$. Also, a Σ-algebra (A, a) satisfies this equation in the classical sense (i.e., for every interpretation $f: \{x_0, \ldots, x_{n-1}\} \to A$, we have $\overline{f}(t) = \overline{f}(t')$) iff the corresponding $\mathcal{T}_\Sigma$-algebra satisfies this equation in the sense of 10.1.

2. In fact, for the theory $\mathcal{T}_\Sigma$, equations in the sense of 10.1 are equivalent to the classical equations: given a parallel pair

$$t, t': n \rightrightarrows k \text{ in } \mathcal{T}_\Sigma$$

and given the k projections $p_i^k: k \to 1$ ($i = 0, \ldots, k-1$) specified by the functor $T_\Sigma: \mathcal{N} \to \mathcal{T}_\Sigma$, we get a k-tuple of equations

$$p_i^k \cdot t = p_i^k \cdot t'$$

in the classical sense. It is clear that a Σ-algebra satisfies each of these k equations iff the corresponding $\mathcal{T}_\Sigma$-algebra satisfies $t = t'$ in the sense of 10.1.

13.10 Definition

1. By an *equational category of Σ-algebras* is meant a full subcategory of Σ-*Alg* formed by all algebras satisfying a set E of equations. We denote such a category by

$$(\Sigma, E)\text{-}Alg.$$

The pair (Σ, E) is called an *equational theory*.

2. *Equational categories* are concrete categories over *Set* that are, for some signature Σ, equational categories of Σ-algebras.

13.11 Theorem *One-sorted algebraic categories are precisely the equational categories. In more detail, a concrete category over Set is one-sorted algebraic iff it is concretely equivalent to an equational category.*

Proof

1. Every equational category of Σ-algebras is a one-sorted algebraic category. In fact, following Remark 13.9, the concrete equivalence Σ-$Alg \simeq Alg\,\mathcal{T}_\Sigma$ of Lemma 13.3 is restricted to a concrete equivalence between (Σ, E)-Alg and $Alg(\mathcal{T}_\Sigma/\sim_E)$, where $\sim_E$ is the congruence on $\mathcal{T}_\Sigma$ generated by E (see 10.7).

2. Conversely, every one-sorted algebraic category is concretely equivalent to an equational category of Σ-algebras. In fact, $Alg\,\mathcal{T}$ is equivalent to $Alg\,(\mathcal{T}_\Sigma/\sim)$ for some congruence $\sim$ on $\mathcal{T}_\Sigma$ (Remark 13.8) and therefore to (Σ, E)-Alg, where E is the set of all equations $u = v$, where u and v are congruent terms. $\square$

13.12 Example Recall that a *semigroup* is an algebra on one associative binary operation. This means that we consider the Σ-algebras with $\Sigma = \{*\}$ that satisfy the equation

$$(x * y) * z = x * (y * z).$$

Thus the theory of semigroups is the quotient theory $\mathcal{T}_\Sigma / \sim$, where $\sim$ is the congruence generated by the preceding equation.

13.13 Example Beside the algebraic theory $\mathcal{T}_{ab}$ of abelian groups of 1.6, we now have a different theory, based on the usual equational presentation: let $\Sigma = \{+, -, 0\}$ with $+$ being binary, $-$ being unary, and 0 being nullary. Then a theory of abelian groups is the quotient $\mathcal{T}_\Sigma / \sim$ modulo the congruence on $\mathcal{T}_\Sigma$ generated by the four equations

$$(x + y) + z = x + (y + z),$$
$$x + y = y + x,$$
$$x + 0 = x,$$
$$x + (-x) = 0.$$

13.14 Example Recall that a *monoid* is a semigroup $(M, *)$ with a unit. We can consider the category of all monoids as the category (Σ, E)-Alg, where Σ has a binary symbol $*$ and a nullary symbol e and E contains the associativity of $*$ and the equations

$$x = x * e$$
$$x = e * x$$

(equivalently, as the category $Alg(\mathcal{T}_\Sigma / \sim)$, where $\sim$ is the congruence generated by the associativity of $*$ and the preceding equations).

13.15 Example For every monoid M (3.10), an *M-set* is a pair (X, α) consisting of a set X and a monoid action $\alpha \colon M \times X \to X$ (the usual notation is mx in place of $\alpha(m, x)$) such that every element $x \in X$ satisfies $m(m'x) = (m * m')x$ for all $m, m' \in M$, and $ex = x$. The homomorphisms $f \colon (X, \alpha) \to (Z, \beta)$ of M-sets are the functions $f \colon X \to Z$ with $f(mx) = mf(x)$ for all $m \in M$ and $x \in X$. We can describe this category as (Σ, E)-Alg, where $\Sigma = M$ with all arities equal to 1 and E consists of the equations

$$x = ex$$
$$(m * m')x = m(m'x)$$

for all $m, m' \in M$.

13.16 Definition A concrete category $U \colon \mathcal{A} \to \mathcal{K}$ is called

1. *amnestic* provided that for every isomorphism $i: A \to A'$ in $\mathcal{A}$ with $Ui = \mathrm{id}_{UA}$ we have $A = A'$ (this implies $i = \mathrm{id}_A$ because U is faithful)
2. *transportable* provided that for every object A in $\mathcal{A}$ and every isomorphism $i: UA \to X$ in $\mathcal{K}$, there exists an isomorphism $j: A \to B$ in $\mathcal{A}$ with $UB = X$ and $Uj = i$
3. *uniquely transportable* if in condition 2 the isomorphism j is unique.

13.17 Example

1. For every one-sorted algebraic theory $(\mathcal{T}, T)$, the concrete category

$$Alg\, T: Alg\, \mathcal{T} \to Set$$

is transportable, but almost never uniquely transportable (see 11.7). In fact, given a $\mathcal{T}$-algebra A and a bijection $i: A1 \to X$, let $B: \mathcal{T} \to Set$ be defined on objects by $Bn = X^n$ and on morphisms $f: n \to k$ in the unique way that makes the powers of i natural:

$$\begin{array}{ccc} (A1)^n \simeq An & \xrightarrow{Af} & Ak = (A1)^k \\ {\scriptstyle i^n}\downarrow & & \downarrow{\scriptstyle i^k} \\ X^n = Bn & \xrightarrow{Bf} & Bk = X^k \end{array}$$

Then these powers form a natural isomorphism $j: A \to B$ with $j_1 = i$.
2. For every signature Σ, the concrete category

$$U_\Sigma: \Sigma\text{-}Alg \to Set, \quad U_\Sigma(A, \sigma^A) = A$$

is uniquely transportable. In fact, given a bijection $i: A \to X$, there is a unique way of defining, for an n-ary symbol $\sigma \in \Sigma$, the operation σ^X so that the square

$$\begin{array}{ccc} A^n & \xrightarrow{\sigma^A} & A \\ {\scriptstyle i^n}\downarrow & & \downarrow{\scriptstyle i} \\ X^n & \xrightarrow{\sigma^X} & X \end{array}$$

commutes. The same is true for the equational categories of Σ-algebras. For example, the category of abelian groups is uniquely transportable.

3. For every endofunctor H of a category $\mathcal{K}$, the concrete category

$$U_H\colon H\text{-}Alg \to \mathcal{K}, \quad (A, a) \mapsto A$$

of 12.1 is uniquely transportable. In fact, given an algebra $a\colon HA \to A$ and an isomorphism $i\colon A \to X$ in $\mathcal{K}$, the unique algebra $x\colon HX \to X$ for which i becomes a homomorphism is given by $x = i \cdot a \cdot Hi^{-1}$.

13.18 Remark

1. For every concrete category, we have

   transportable + amnestic $\Leftrightarrow$ uniquely transportable.

   In fact, if $(\mathcal{A}, U)$ is transportable and amnestic, and if in Definition 13.16.2 we have another isomorphism $j'\colon A \to B'$ with $Uj' = i$, use amnesticity on the isomorphism $j' \cdot j^{-1}\colon B \to B'$ to conclude $B = B'$. Then $j = j'$ since U is faithful. Conversely, if $(\mathcal{A}, U)$ is uniquely transportable, then by applying 13.16.3 to $i = \mathrm{id}_{UA}$, we deduce that it is amnestic.
2. Transportability is not invariant under concrete equivalence and thus not all one-sorted algebraic categories are transportable. For example, let $E\colon Set' \to Set$ be the full subcategory of Set consisting of all cardinal numbers. Then (Set', E) is one-sorted algebraic because it is concretely equivalent to (Set, Id), but it obviously fails to be transportable.
3. We will see in Corollary 13.21 a converse of Example 13.17.2: every uniquely transportable one-sorted algebraic category is (up to concrete isomorphism) an equational category.

13.19 Definition   Given concrete categories $U\colon \mathcal{A} \to \mathcal{K}$ and $V\colon \mathcal{B} \to \mathcal{K}$, by a *concrete isomorphism* between them we mean a concrete functor

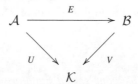

for which there exists a functor $E'\colon \mathcal{B} \to \mathcal{A}$ such that both $E \cdot E'$ and $E' \cdot E$ are equal to the identity functors. (Note that such a functor E' is necessarily concrete.) We then say that $(\mathcal{A}, U)$ and $(\mathcal{B}, V)$ are *concretely isomorphic*.

13.20 Lemma   *A concrete equivalence between uniquely transportable concrete categories is a concrete isomorphism.*

Proof Given a concrete equivalence

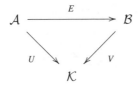

between uniquely transportable categories, we prove that E is bijective on objects – thus it is a (concrete) isomorphism.

1. If A and A' are objects of $\mathcal{A}$ with $EA = EA'$, then for the identity morphism of EA, there exists, since E is full, a morphism $f: A \to A'$ with $Ef = \mathrm{id}$. And f is of course an isomorphism in $\mathcal{A}$. Since $Uf = V(Ef) = \mathrm{id}$, amnesticity of U implies $A = A'$.

2. For every object B of $\mathcal{B}$, there exists an isomorphism $i: EA \to B$ in $\mathcal{B}$ yielding an isomorphism $Vi: UA \to VB$ in $\mathcal{K}$. Let $j: A \to A'$ be the unique isomorphism in $\mathcal{A}$ with $Uj = Vi$. The isomorphism $Ej \cdot i^{-1}: B \to EA'$ fulfils $V(Ej \cdot i^{-1}) = Uj \cdot Vi^{-1} = \mathrm{id}$, and thus by amnesticity of V, we have $B = EA'$. □

13.21 Corollary *Uniquely transportable one-sorted algebraic categories are, up to concrete isomorphism, precisely the equational categories.*

In fact, this follows from Theorem 13.11, Example 13.17, and Lemma 13.20.

Birkhoff's variety theorem (10.22) can be restated in the context of Σ-algebras:

13.22 Theorem *Let Σ be a signature. A full subcategory $\mathcal{A}$ of Σ-Alg is equational iff it is closed in Σ-Alg under*

1. *products*
2. *subalgebras and*
3. *regular quotients*

In fact, this follows from 10.22, 11.34, and 13.11.

13.23 Proposition *Let Σ be a one-sorted signature and $H: \mathrm{Set} \to \mathrm{Set}$ a quotient of the polynomial functor H_Σ. The concrete category H-Alg is concretely isomorphic to an equational category of Σ-algebras.*

Proof Let $\alpha: H_\Sigma \to H$ be a natural transformation with epimorphic components. We get a full and faithful functor

$$I: H\text{-}Alg \to H_\Sigma\text{-}Alg \text{ defined by } I(A, a) = (A, a \cdot \alpha_A).$$

Moreover, I is concrete since the diagram

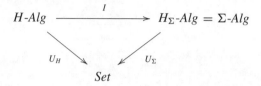

clearly commutes.

Since I is injective on objects, H-Alg is concretely isomorphic to the full subcategory $I(H$-$Alg)$ of Σ-Alg. We are to prove that $I(H$-$Alg)$ satisfies the conditions of Theorem 13.22.

1. Consider the commutative diagram above. Since products are both preserved by U_H and reflected by U_Σ, they are also preserved by I. In particular, $I(H$-$Alg)$ is closed in Σ-Alg under products.

2. Let $f: (A, x) \to I(B, b)$ be a monomorphism in Σ-Alg (and then in Set). Since α_A is a strong epimorphism in Set, we get an H-algebra structure on A by diagonal fill-in:

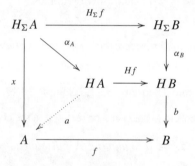

This shows that $(A, x) = I(A, a)$.

3. First observe that in point 2 if $f: I(A, a) \to (B, y)$ is an isomorphism in Σ-Alg, then $(B, y) \in I(H$-$Alg)$. Indeed $f: (A, a) \to (B, b)$ is an isomorphism in H-Alg, where $b = f \cdot a \cdot Hf^{-1}$ and $I(B, b) = (B, y)$. (In other words, $(H$-$Alg, I)$ is transportable; see Definition 13.16.)

Now consider a regular epimorphism $e: I(A, a) \to (B, y)$ in Σ-Alg. Its kernel pair, being a subobject of $I(A, a) \times I(A, a)$, lies in the image of I, and we can take its coequalizer (Q, q) in H-Alg. This coequalizer is preserved by I (because by 3.4 it is reflexive, and I preserves sifted colimits: this is so because sifted colimits are both preserved by U_H and reflected by U_Σ). Thus $I(Q, q) \simeq (B, y)$. □

13.24 Remark In general algebra, the concepts of finitely generated and finitely presentable Σ-algebra are defined as follows: a Σ-algebra (A, a) is called

1. finitely generated if it is generated by a finite set (see 11.10); that is, (A, a) is isomorphic to

$$F_\Sigma\{x_1, \ldots, x_n\}/\sim$$

for some congruence $\sim$ on $F_\Sigma\{x_1, \ldots, x_n\}$
2. finitely presentable if it is such a quotient modulo a finitely generated congruence; that is, (A, a) is isomorphic to $F_\Sigma\{x_1, \ldots, x_n\}/\sim$ for some congruence $\sim$ generated by finitely many equations

It turns out that these concepts coincide with the categorical concepts of 5.21 and 5.3, respectively. Let us first observe the following:

1. Every subalgebra generated by a set X (see 11.10) is a regular quotient of the free algebra $F_\Sigma X$. In fact, let B be the subalgebra of A generated by X, and let $f: F_\Sigma X \to A$ be the unique homomorphism extending the inclusion map. Then the image $f(F_\Sigma X)$ is a subalgebra of A because the forgetful functor preserves regular factorizations (see 11.9); this is, obviously, the least subalgebra containing X, and the codomain restriction of f is a regular epimorphism.
2. Conversely, every regular quotient $q: F_\Sigma X \to A$ of a free algebra generated by X is "generated by X" — more precisely, the image of the map

$$X \xrightarrow{\eta_\Sigma} U_\Sigma(F_\Sigma X) \xrightarrow{U_\Sigma q} U_\Sigma A$$

generates A.

13.25 Proposition *A Σ-algebra is a finitely generated object of Σ-Alg iff it is a regular quotient of a finitely generated free algebra.*

Proof
1. Let (A, a) be a finitely generated object of Σ-*Alg*. Form a diagram in Σ-*Alg* indexed by the poset of all finite subsets of A by assigning to every such $X \subseteq A$ the subalgebra $\overline{X}$ of A generated by X (see 11.10). Given finite subsets X and Z with $X \subseteq Z \subseteq A$, the connecting map $\overline{X} \to \overline{Z}$ is the inclusion map. Then the inclusion homomorphisms $i_X: \overline{X} \to A$ form a colimit cocone of this directed diagram. Since the functor Σ-*Alg*$(A, -)$ preserves this colimit, for $\mathrm{id}_A \in \Sigma$-*Alg*(A, A) there exists a finite set X such that id_A lies in the image of i_X – but this proves $\overline{X} = A$.

2. Let us prove that every quotient $A = F\{x_1, \ldots, x_n\}/\sim$ is finitely generated in the sense of 11.10. Given a directed diagram of subobjects B_i $(i \in I)$ with a colimit $B = \mathrm{colim}\, B_i$, it is our task to prove that Σ-*Alg*$(A, -)$ preserves this colimit; that is, every homomorphism $h: A \to B$ factorizes through one

of the colimit homomorphisms $b_i \colon B_i \to B$. For the finite set $\{h(\eta_X(x_k))\}_{k=1}^n$, there exists $i \in I$ such that this set lies in the image of b_i. From that it easily follows that the image of h is contained in the image of b_i. Since b_i is a monomorphism, it follows that there exists a homomorphism $g \colon A \to B_i$ with $h = b_i \cdot g$, as requested. $\square$

13.26 Proposition *A Σ-algebra is a finitely presentable object of Σ-Alg iff it is a regular quotient of a finitely generated free algebra modulo a finitely generated congruence.*

Proof
1. Let (A, a) be a finitely presentable object. By 11.28, there exists a coequalizer

$$F_\Sigma X \underset{v}{\overset{u}{\rightrightarrows}} F_\Sigma Z \xrightarrow{c} (A, a)$$

with X and Z finite. Let $\sim$ be the congruence generated by the finitely many equations $u(\eta_\Sigma(x)) = v(\eta_\Sigma(x))$ where $x \in X$; then (A, a) is clearly isomorphic to $F_\Sigma Z / \sim$: the canonical morphism $q \colon F_\Sigma Z \to F_\Sigma Z / \sim$ is, namely, also a coequalizer of u and v.

2. Conversely, let Z be a finite set and $\sim$ a congruence on $F_\Sigma Z$ generated by equations $t_1 = s_1, \ldots, t_k = s_k$. For $X = \{1, \ldots, k\}$, define homomorphisms $u, v \colon F_\Sigma X \rightrightarrows F_\Sigma Z$ by

$$u(i) = t_i \text{ and } v(i) = s_i \text{ for } i = 1, \ldots, k.$$

Then the canonical map $q \colon F_\Sigma Z \to F_\Sigma Z / \sim$ is a coequalizer of u and v. Therefore $F_\Sigma Z / \sim$ is finitely presentable by 11.28. $\square$

Historical remarks

The material of this chapter is closely related to the classical work of Birkhoff (1935); for a modern exposition of general algebra, see, for example, Cohn (1965) or Grätzer (2008). The concepts of (uniquely) transportable and amnestic functors are taken from Adámek et al. (2009); some authors use transportability requesting uniqueness as a part of the definition.

14
S-sorted algebraic categories

In previous chapters, we have considered one-sorted algebraic categories, which are categories equipped with a forgetful functor into *Set*, such as groups, abelian groups, and lattices. In computer science, one often considers S-sorted algebras, where S is a given nonempty set (of sorts), and algebras are not sets with operations but rather S-indexed families of sets with operations of given sort. This means that the forgetful functor is into Set^S rather than into *Set*. In this chapter, we revisit one-sorted algebraic categories generalizing definitions and several results to the S-sorted case.

Analogously to the one-sorted case, where the theory has objects X^n (which we represented by n alone) and projections $\pi_i^n \colon X^n \to X$ are specified, in the case of S-sorted theories, we have objects X_s for $s \in S$ that generate the whole theory in the sense that every object of $\mathcal{T}$ is a product

$$X_{s_0} \times \ldots \times X_{s_{n-1}}$$

for some word $w = s_0 \ldots s_{n-1}$ over S. We again suppose that projections

$$\pi_i^w \colon X_{s_0} \times \ldots \times X_{s_{n-1}} \to X_{s_i} \quad (i = 0, \ldots, n-1)$$

are chosen, and again, instead of working with the preceding product, we work with the word $s_0 \ldots s_{n-1}$ alone. In other words, the theory $\mathcal{N}$ that plays a central role for one-sorted theories is generalized to the following.

14.1 Notation Recall from 1.5 that we denote by

$$S^*$$

the category whose objects are the finite words over S and whose morphisms from $s_0 \ldots s_{n-1}$ to $t_0 \ldots t_{k-1}$ are all functions $f \colon k \to n$ with $s_{f(i)} = t_i$ for all

$i = 0, \ldots, k-1$. In particular, for every word $w = s_0 \ldots s_{n-1}$, we have the projections

$$\pi_i^w : s_0 \ldots s_{n-1} \to s_i \quad (i = 0, \ldots, n-1)$$

given by the ith injection $1 \mapsto n$ in *Set*.

14.2 Example $\mathcal{N} = \{s\}^*$ provided that we identify every natural number n with the word $ss \ldots s$ of length n.

14.3 Remark We know from 1.5 that S^* is an algebraic theory for Set^S, and every word w is a product of one-letter words with the projections $\pi_0^w, \ldots, \pi_{n-1}^w$ given earlier. We are going to identify Set^S with $Alg\, S^*$. The full embedding

$$Y_{S^*} : (S^*)^{op} \to Set^S$$

assigns to a word $w = s_0 \ldots s_{n-1}$ the S-sorted set

$$Y_{S^*}(w)_s = \{i = 0, \ldots, n-1 ; \ s_i = s\}.$$

14.4 Definition Let S be a nonempty set.

1. An *S-sorted algebraic theory* is a pair $(\mathcal{T}, T)$ where $\mathcal{T}$ is an algebraic theory whose objects are the words over S, and $T : S^* \to \mathcal{T}$ is a theory morphism that is the identity map on objects.
2. A *morphism* of S-sorted algebraic theories $M : (\mathcal{T}_1, T_1) \to (\mathcal{T}_2, T_2)$ is a functor $M : \mathcal{T}_1 \to \mathcal{T}_2$ such that $M \cdot T_1 = T_2$:

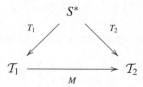

14.5 Remark Analogously to 11.3, we have not requested that morphisms of one sorted theories preserve finite products since this simply follows from the equation $M \cdot T_1 = T_2$. Observe that this equation implies that M is the identity map on objects.

14.6 Example

1. For the category *Graph* of graphs, see 1.11, we have an S-sorted theory with $S = \{v, e\}$, and $T : S^* \to \mathcal{T}_{graph}$ is determined by $Tv = v$ and $Te = e$ (the theory $\mathcal{T}_{graph}$ is described in 1.16).
2. Let $\mathcal{C}$ be a small category, put $S = obj\, \mathcal{C}$, and let $E_{Th} : \mathcal{C} \to \mathcal{T}_\mathcal{C}$ denote the free completion of $\mathcal{C}$ under finite products (1.14); recall from 1.15 that

the objects of $\mathcal{T}_C$ can be viewed as words over S. Therefore we have a unique theory morphism $T_C \colon S^* \to \mathcal{T}_C$ that is the identity map on objects. We obtain an S-sorted theory

$$(\mathcal{T}_C, T_C).$$

14.7 Remark Precisely as in the one-sorted case, the functor T does not influence the concept of algebra: the category $Alg\,\mathcal{T}$ thus consists, again, of all functors $A \colon \mathcal{T} \to Set$ preserving finite products. However, the presence of T makes the category of algebras concrete over Set^S: the forgetful functor is simply

$$Alg\,T \colon Alg\,\mathcal{T} \to Set^S$$

(see Definition 14.4). More precisely, this forgetful functor takes an algebra $A \colon \mathcal{T} \to Set$ to the S-sorted set $\langle As \rangle_{s \in S}$ and a homomorphism $h \colon A \to B$ to the S-sorted function with components $h_s \colon As \to Bs$.

14.8 Proposition *Let $(\mathcal{T}, T)$ be an S-sorted algebraic theory. The forgetful functor*

$$Alg\,T \colon Alg\,\mathcal{T} \to Set^S$$

is faithful, algebraic, and conservative. It thus preserves and reflects limits, sifted colimits, monomorphisms, and regular epimorphisms.

The proof is analogous to that of 11.8.

14.9 Remark The concept of one-sorted algebraic category of Chapter 11 used concrete equivalences of categories over Set. For S-sorted algebraic theories, we need, analogously, concrete equivalences over Set^S (see 11.12).

14.10 Definition An *S-sorted algebraic category* is a concrete category over Set^S that is concretely equivalent to $Alg\,T \colon Alg\,\mathcal{T} \to Set^S$ for an S-sorted algebraic theory $(\mathcal{T}, T)$.

14.11 Proposition *Every variety of T-algebras for an S-sorted theory $(\mathcal{T}, T)$ is an S-sorted algebraic category.*

The proof is analogous to that of 11.17.

14.12 Remark Let $(\mathcal{T}, T)$ be an S-sorted theory.

1. The forgetful functor $Alg\,T \colon Alg\,\mathcal{T} \to Set^S$ has a left adjoint. In fact, due to 4.11 being applied to $Y_{S^*} \colon (S^*)^{op} \to Set^S$, we can choose a left adjoint

$$F_T \colon Set^S \to Alg\,\mathcal{T}$$

in such a way that the square

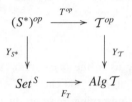

commutes.

2. $\mathcal{T}$-algebras of the form $F_\mathcal{T}(X)$, for X an S-sorted set, are called *free algebras*. If X is finite (1.18), they are called *finitely generated free algebras*.

14.13 Corollary *Let $(\mathcal{T}, T)$ be an S-sorted theory. $\mathcal{T}^{op}$ is equivalent to the full subcategory of Alg $\mathcal{T}$ of finitely generated free algebras.*

14.14 Remark Results about finitely presentable and perfectly presentable algebras and regular projectives generalize easily from the one-sorted case (see 11.26–11.33), to the S-sorted case; we leave this for the reader. Let us just stress that when working with variables in the S-sorted case, a sort is assigned to every variable; that is, the corresponding object of variables also lives in Set^S.

14.15 Theorem: *S-sorted algebraic duality* *The 2-category ALG ^{S} of S-sorted algebraic categories, concrete functors, and natural transformations is biequivalent to the dual of the 2-category Th ^{S} of S-sorted algebraic theories, morphisms of S-sorted theories, and natural transformations.*

The proof is completely analogous to that of 11.38.

14.16 Definition Let Σ be an S-sorted signature, see Example 11.10.

1. A *Σ-algebra* is a pair (A, a) consisting of an S-sorted set $A = \langle A_s \rangle_{s \in S}$ and a function a assigning to every element $\sigma: s_0 \ldots s_{n-1} \to s$ of Σ a mapping

$$\sigma^A: A_{s_0} \times \ldots \times A_{s_{n-1}} \to A_s.$$

(In case $n = 0$, we have a constant $\sigma^A \in A_s$.)

2. *Σ-homomorphisms* from (A, a) to (B, b) are S-sorted functions

$$f = \langle f_s \rangle \text{ with } f_s: A_s \to B_s \ (s \in S)$$

such that for every operation $\sigma: s_0 \ldots s_{n-1} \to s$ of Σ, the square

$$\begin{array}{ccc} A_{s_0} \times \ldots \times A_{s_{n-1}} & \xrightarrow{\sigma^A} & A_s \\ {\scriptstyle f_{s_0} \times \ldots \times f_{s_{n-1}}} \downarrow & & \downarrow {\scriptstyle f_s} \\ B_{s_0} \times \ldots \times B_{s_{n-1}} & \xrightarrow{\sigma^B} & B_s \end{array}$$

commutes. This yields a concrete category

$$\Sigma\text{-}Alg$$

of Σ-algebras with the forgetful functor

$$U_\Sigma: \Sigma\text{-}Alg \to Set^S, \quad U_\Sigma(A, a) = A.$$

14.17 Example

1. The category of graphs has the form Σ-Alg for $S = \{v, e\}$ and Σ consisting of two operations of arity $e \to v$ (called τ and σ in 1.11).
2. For $\Sigma = \emptyset$, we have Σ-$Alg = Set^S$.
3. For sequential automata (see 1.25), put $S = \{s, i, o\}$ and $\Sigma = \{\delta, \gamma, \varphi\}$ with arities $\delta: si \to s$, $\gamma: s \to o$ and $\varphi: s$.
4. For the example of stacks 1.24, put $S = \{s, n\}$ and $\Sigma = \{\text{succ}, \text{push}, \text{pop}, \text{top}, 0, e\}$ with the arities given in 1.24.

14.18 Remark

1. The description of a left adjoint

$$F_\Sigma: Set^S \to \Sigma\text{-}Alg$$

of U_Σ is completely analogous to 13.1. Given an S-sorted set X of variables, we form the smallest S-sorted set $F_\Sigma X$ (of terms) such that every element $x \in X_s$ is a term of sort s, and for every $\sigma \in \Sigma$ of arity $s_0 \ldots s_{n-1} \to s$ and for every n-tuple of terms $p_0, \ldots, p_{n-1}$ of sorts $s_0, \ldots, s_{n-1}$, respectively, we have a term $\sigma(p_0, \ldots, p_{n-1})$ of sort s. The Σ-algebra structure on $F_\Sigma X$ is given by the formation of terms $\sigma(p_0, \ldots, p_{n-1})$.

   This defines a functor $F_\Sigma: Set^S \to \Sigma\text{-}Alg$ on objects. To define it on morphisms, proceed as in 13.1.

2. We obtain, assuming a countable set of "standard variables x_i^s of sort s" for every $s \in S$, an S-sorted theory

$$(\mathcal{T}_\Sigma, T_\Sigma)$$

analogous to the one described in 13.2: the words $s_0 \ldots s_{n-1}$ represent the free Σ-algebra $F_\Sigma\{x_0^{s_0}, \ldots, x_{n-1}^{s_{n-1}}\}$. The categories Σ-Alg and $Alg\,T_\Sigma$ are concretely equivalent over Set^S; this is analogous to 13.3.

3. Equations in the sense of 10.1 can be substituted by expressions

$$t = t',$$

where t and t' are two elements of $F_\Sigma X$ of the same sort (for some finite S-sorted set X of standard variables). This is analogous to 13.9, except that in the S-sorted case, the quantification of variables must be made explicit. If Σ is a one-sorted signature and t, t' are terms in $F_\Sigma X$, then in place of X, we can take the set $Z \subseteq X$ of all variables that appear in t or t'. An algebra satisfies $t = t'$ independently of whether we work with $F_\Sigma Z$ or $F_\Sigma X$. This is not so in S-sorted signatures, as we demonstrate in 14.20.1. We therefore need the following.

14.19 Definition Given an S-sorted signature Σ, by an *equation* is meant an expression

$$\forall x_0 \, \forall x_1 \ldots \forall x_{n-1} \, (t = t'),$$

where x_i is a variable of sort s_i ($i = 0, \ldots, n-1$) and t, t' are elements of $F_\Sigma\{x_0, \ldots, x_{n-1}\}$ of the same sort s. A Σ-algebra (A, a) *satisfies* the equation provided that for every S-sorted function $f: \{x_0, \ldots, x_{n-1}\} \to A$, the unique homomorphism $\overline{f}: F_\Sigma\{x_0, \ldots, x_{n-1}\} \to (A, a)$ extending f fulfils $\overline{f}_s(t) = \overline{f}_s(t')$. In case $n = 0$, we write $\forall \emptyset \, (t = t')$.

14.20 Example

1. In the signature of graphs (see 14.17), consider variables x, x' of sort v and a variable y of sort e. The equation

$$\forall x \, \forall x' \, (x = x')$$

describes graphs on at most one vertex, whereas

$$\forall x \, \forall x' \, \forall y \, (x = x')$$

describes all graphs that either have no edge or have just one vertex.

2. In the signature of stacks, (see 14.17) there are several equations one expects to be required. For example, if a natural number x is inserted into a stack y and then deleted, the stack does not change:

$$\forall x \, \forall y \, (\text{pop}(\text{push}(x, y)) = y).$$

Other such equations are

$$\forall x\, \forall y\, (\text{top}(\text{push}(x, y))) = x$$

and (from our definition of *top*)

$$\text{top}(e) = 0.$$

14.21 Example Let us return to Example 10.23, explaining that Birkhoff's variety theorem requires, in general, the use of directed unions. The example worked with $Set^{\mathbb{N}}$, which is Σ-*Alg* for the empty $\mathbb{N}$-sorted signature. Let x_n and y_n be variables of sort $n \in \mathbb{N}$, and consider the equation quantifying y_n and all $x_0, x_1, x_2, \ldots$:

$$\forall y_n\, \forall x_0\, \forall x_1\, \forall x_2 \ldots (x_n = y_n).$$

Then algebras, that is, $\mathbb{N}$-sorted sets, satisfy these equations iff they lie in the category $\mathcal{A}$. However, infinite quantification brings us out of the finitary logic (and out of the realm of Definition 10.1).

14.22 Definition

1. Let Σ be an S-sorted signature. By an *S-sorted equational category of Σ-algebras* is meant a full subcategory of Σ-*Alg* formed by all algebras satisfying a set E of equations (in the sense of 10.1 or, equivalently, Remark 14.18).
2. *S-sorted equational categories* are concrete categories over Set^S that are, for some signature Σ, S-sorted equational categories of Σ-algebras.

14.23 Remark Birkhoff's variety theorem for S-sorted algebras states that S-sorted equational categories are precisely the full subcategories of Σ-*Alg* closed under

products
subalgebras
regular quotients
directed unions

If S is a finite set, the last item can be left out. The proof is completely analogous to that of 11.34: we choose i_0 in such a way that given $s \in S$ for which the sort of A is nonempty, the sort of A_{i_0} is also nonempty.

14.24 Example All one-sorted theories form a many-sorted equational category. In more detail, the category Th^1 of one-sorted theories and their morphisms is an equational category of Σ-algebras for the following signature using

$\mathbb{N} \times \mathbb{N}$ as a set of sorts: Σ consists of the binary operations "composition,"

$$c_{ijk}: (i, j)(j, k) \to (i, k) \quad \text{(for all } i, j, k \in \mathbb{N}\text{);}$$

the constants expressing identity morphisms,

$$e_n: (n, n) \quad \text{(for all } n \in \mathbb{N}\text{);}$$

and the projections (provided by $T: \mathcal{N} \to \mathcal{T}$),

$$p_{n,k}: (n, k) \quad \text{(for all } k < n\text{).}$$

In fact, with every one-sorted theory $(\mathcal{T}, T)$, associate the Σ-algebra $\overline{(\mathcal{T}, T)}$ whose underlying sets are the hom-sets of $\mathcal{T}$,

$$\overline{(\mathcal{T}, T)}_{(i,j)} = \mathcal{T}(i, j),$$

and where c_{ijk} and e_n have the obvious meaning, and the interpretation of $p_{n,k}$ is $T\pi_k^n$. Every morphism $F: (\mathcal{T}, T) \to (\mathcal{T}', T')$ of one-sorted theories defines a homomorphism of Σ-algebras $\overline{F}: \overline{(\mathcal{T}, T)} \to \overline{(\mathcal{T}', T')}$ whose underlying function of sort (i, j) is the action of F at $\mathcal{T}(i, j)$. Then

$$\overline{(-)}: Th^1 \to \Sigma\text{-}Alg$$

is a full and faithful functor. The image of this full embedding is the equational category described by the equations expressing the fact that e_n is the identity morphism

$$c_{ijj}(f, e_j) = f \quad \text{and} \quad c_{iij}(e_i, g) = g$$

and the associativity of composition

$$c_{ikn}(c_{ijk}(f, g), h) = c_{ijn}(f, c_{jkn}(g, h)).$$

14.25 Remark

1. Another way of expressing Th^1 as an equational category of many-sorted algebras uses only $\mathbb{N}$ as the set of sorts: the underlying sets of the algebra for $(\mathcal{T}, T)$ are then $\mathcal{T}(n, 1)$ for $n \in \mathbb{N}$. Here, however, the corresponding operations need to take into account the tupling $g(f_1, \ldots, f_n)$ of an n-ary operation symbol $g \in \mathcal{T}(n, 1)$ with n symbols $f_i \in \mathcal{T}(k_i, 1)$: put

$$k = k_1 + \ldots + k_n;$$

then our signature must contain "tupling" operation symbols

$$t_{k_1 \ldots k_n}: k_1 \ldots k_n \to k$$

(for all n-tuples $k_1, \ldots, k_n$ in $\mathbb{N}$). The equations describing Th^1 as an equational category of $\mathbb{N}$-sorted algebras are then somewhat more involved.

2. The presentation of one-sorted theories as an equational category of $\mathbb{N}$-sorted algebras above is closely related to the concept of clone: given a Σ-algebra A on a set X, the *clone* of A is the smallest set of functions of many variables $f: X^n \to X$ ($n \in \mathbb{N}$) containing all projections and all operations σ^A for $\sigma \in \Sigma$ and closed under the tupling $g(f_1, \ldots, f_n)$. One then introduces a partial operation of (simultaneous) composition on the clone.

14.26 Example All S-sorted theories form a many-sorted equational catgeory. This is completely analogous to the previous one-sorted case. A trivial presentation of Th^S uses sorts $S^* \times S^*$ (interpreted as the hom-sets of the theory). A more "clonelike" presentation uses sorts $S^* \times S$ since the hom-sets $T(w, s)$ for $w \in S^*$ and $s \in S$ are sufficient.

14.27 Example: Modules over variable rings Here we consider pairs (R, M) where M is a right R-module as objects. Morphisms from (R, M) to (R', M') are pairs of functions (h, f), where $h: R \to R'$ is a ring homomorphism and $f: M \to M'$ is a homomorphism of abelian groups satisfying

$$f(\lambda x) = h(\lambda) f(x) \quad \text{for all } \lambda \in R, x \in M.$$

This is an equational category of two-sorted algebras of sorts r and m with ring operations

$+: rr \to r,$
$-: r \to r,$
$\times: rr \to r,$
$0, 1: r,$

and module operations

$\oplus: mm \to m,$
$\ominus: m \to m,$
$\odot: m,$
$*: rm \to m,$

satisfying the equations (1) of the presentation of rings; (2) of abelian groups for $\oplus$, $\ominus$, and $\odot$; and (3) the distributive laws.

14.28 Proposition

1. *S-sorted algebraic categories are precisely the S-sorted equational categories. In more detail, a concrete category over Set^S is S-sorted algebraic iff it is concretely equivalent to an S-sorted equational category of Σ-algebras for some signature Σ.*

2. Uniquely transportable S-sorted algebraic categories are, up to concrete isomorphism, precisely the S-sorted equational categories.

The proofs are completely analogous to those of 13.11 and 13.21.

14.29 Remark Every S-sorted equational category is, of course, complete and cocomplete. In particular, initial algebras exist in all S-sorted equational categories. In theoretical computer science, these algebras are used as a formalization of "abstract data types": these are given by operations and equations and consist of elements generated by the given operations (no extra variables are used), and they satisfy only the equations that are consequences of the given ones. An abstract data type is thus, precisely as initial objects should be, determined only up to isomorphism. We illustrate this with a couple of examples.

14.30 Example

1. Natural numbers form a one-sorted abstract data type given by a constant 0 and a unary operation s (successor). This corresponds to the initial algebra of the one-sorted signature $\Sigma = \{s, 0\}$ with arity 1 and 0, respectively. In fact, every initial Σ-algebra is a representation of natural numbers.
2. For stacks of natural numbers, we need the two-sorted signature Σ of Example 14.17.4. Its initial algebra does not resemble stacks because we will have formal terms such as top(e), top(top(e)), and so on. However, the equational category given by the three equations of Example 14.20.2 has an initial algebra $I = \langle I_n, I_s \rangle$, where I_n is the abstract data type of natural numbers (no equation involves the operation succ) and I_s consists of stacks

$$e = [\ ], [x], [x, y], [x, y, z], \ldots$$

of elements $x, y, z, \ldots$ of I_n.

14.31 Remark For one-sorted signatures, we have Σ-Alg concretely equivalent to H_Σ-Alg, where $H_\Sigma A = \coprod_{\sigma \in \Sigma} A^n$ (for $n =$ arity of σ), (see 2.25).

Analogously for S-sorted signatures Σ; define

$$H_\Sigma \colon Set^S \to Set^S$$

on objects $A = \langle A_s \rangle_{s \in S}$ by setting the sort s of $H_\Sigma A$ as follows: we denote by $\Sigma_s \subseteq \Sigma$ the set of all symbols of output sort s and put

$$(H_\Sigma A)_s = \coprod_{\sigma \in \Sigma_s} A_{s_0} \times \ldots \times A_{s_{n-1}}$$

for the arity $s_0 \ldots s_{n-1} \to s$ of σ. Then there is a concrete equivalence

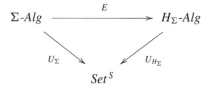

assigning to every Σ-algebra (A, a) the H_Σ-algebra $(A, \bar{a})$ where the coproduct components of

$$\bar{a}_s: \coprod_{\sigma \in \Sigma_s} A_{s_0} \times \ldots \times A_{s_{n-1}} \to A_s$$

are the given operations σ^A.

14.32 Proposition *Every finitary endofunctor H of Set^S is a quotient of a polynomial functor H_Σ for some S-sorted signature Σ. Moreover, the concrete category H-Alg is concretely isomorphic to an equational category of Σ-algebras.*

Proof For the first statement, the argument in 12.15 using the Yoneda lemma generalizes without a problem: define an S-sorted signature Σ whose operations σ of (an arbitrary) arity $s_0 \ldots s_{n-1} \to s$ are precisely the elements of sort s in HX, where the S-sorted set X is given by

$$X_t = \{i = 0, \ldots, n-1; s_i = t\} \text{ for all } t \in S.$$

Then define $\alpha \colon H_\Sigma \to H$ by taking such an operation symbol σ and putting

$$\alpha_Z(\sigma(f)) = (Hf)_s(\sigma)$$

for all S-sorted functions $f \colon X \to Z$.

For the second statement, the only difference to the proof of 13.23 is that in the S-sorted case, we must also check that $I(H\text{-}Alg)$ is closed in Σ-Alg under directed unions. This follows from the commutativity of the diagram

$$\begin{array}{ccc} H\text{-}Alg & \xrightarrow{I} & H_\Sigma\text{-}Alg = \Sigma\text{-}Alg \\ & \searrow U_H \quad \swarrow U_\Sigma & \\ & Set^S & \end{array}$$

since U_H preserves sifted colimits and U_Σ reflects them (see 12.17.3). □

14.33 Remark

1. The converse implication of Theorem 12.15 does not generalize to the S-sorted case: a quotient of a polynomial functor on Set^S need not be finitary.

   A simple example can be presented in $Set^{\mathbb{N}}$: start with the constant functor of value $2 = 1 + 1$ (the S-sorted set having two elements in every sort). This functor is clearly polynomial. Let H be the quotient with $HX = 1$ whenever all sorts of X are nonempty, else $HX = 2$. This functor does not preserve the filtered colimit of all finitely presentable subobjects of 1.

2. For finite sets S of sorts, Theorem 12.15 fully generalizes; finitary endofunctors of Set^S are precisely the quotients of polynomial functors. In fact, the proof of 12.15 is easily modified: in part (b) of the implication $3 \Rightarrow 1$, choose the S-sorted set Z' in such a way that for every sort s, we have $Z'_s \neq \emptyset$ iff $C_s \neq \emptyset$; since $\mathcal{D}$ is filtered and S is finite, this choice is clearly possible. Then, again, $c'_d \cdot i'$ is a split monomorphism.

Historical remarks

Historical comments on S-sorted algebras are mentioned at the end of Chapter 1. For a short introduction to applications of S-sorted algebras, see Wechler (1992).

PART III

Special topics

Modern algebra also enables one to reinterpret the results of classical algebra, giving them far greater unity and generality.
– G. Birkhoff and S. Mac Lane, *A Survey of Modern Algebra,*
Macmillan, New York, 1965: v

15
Morita equivalence

In this chapter, we study the problem of the presentation of an algebraic category by different algebraic theories. This is inspired by the classical work of Kiiti Morita, who, in the 1950s, studied this problem for the categories *R-Mod* of left modules over a ring *R*. He completely characterized pairs of rings *R* and *S* such that *R-Mod* and *S-Mod* are equivalent categories; such rings are nowadays said to be *Morita equivalent*. We will recall the results of Morita subsequently, and we will show in which way they generalize from *R-Mod* to *Alg T*, where *T* is an algebraic theory. We begin with a particularly simple example.

15.1 Example In 1.4, we described a one-sorted algebraic theory $\mathcal{N}$ of *Set*: $\mathcal{N}$ is the full subcategory of Set^{op} whose objects are the natural numbers. Here is another one-sorted theory of *Set*: $\mathcal{T}_2$ is the full subcategory of Set^{op} whose objects are the even natural numbers $0, 2, 4, 6, \ldots$. $\mathcal{T}_2$ obviously has finite products. Observe that $\mathcal{T}_2$ is not idempotent complete (consider the constant functions $2 \to 2$) and that $\mathcal{N}$ is an idempotent completion of $\mathcal{T}_2$: for every natural number n, we can find an idempotent function $f: 2n \to 2n$ with precisely n fixed points. Then n is obtained by splitting f. Following 6.14 and 8.12, $Alg\, \mathcal{T}_2 \simeq Alg\, \mathcal{N} \simeq Set$.

In fact, we can repeat the previous argument for every natural number $k > 0$. In this way, we get a family $\mathcal{T}_k$, $k = 1, 2, \ldots$ of one-sorted algebraic theories of *Set* (with $\mathcal{T}_1 = \mathcal{N}$). We will prove later that up to equivalence, there is no other one-sorted algebraic theory of *Set*.

Clearly, if $\mathcal{T}$ and $\mathcal{T}'$ are algebraic theories and if there is an equivalence $\mathcal{T} \simeq \mathcal{T}'$, then $Alg\, \mathcal{T}$ and $Alg\, \mathcal{T}'$ are equivalent categories. The previous example shows that the converse is not true.

15.2 Definition Two algebraic theories T and T' are called *Morita equivalent* if the corresponding categories $Alg\,T$ and $Alg\,T'$ are equivalent.

From 6.14 and 8.12, we already know a simple characterization of Morita-equivalent algebraic theories: two theories are Morita equivalent iff they have equivalent idempotent completions. In the case of S-sorted algebraic categories, a much sharper result can be proved. Before doing so, let us recall the classical result of Morita.

15.3 Example Let R be a unitary ring (not necessarily commutative) and denote by R-*Mod* the category of left R-modules. There are two basic constructions:

1. *Matrix ring* $R^{[k]}$. This is the ring of all $k \times k$ matrices over R with the usual addition, multiplication, and unit matrix. This ring $R^{[k]}$ is Morita equivalent to R for every $k > 0$; that is, the category $R^{[k]}$-*Mod* is equivalent to R-*Mod*.
2. *Idempotent modification* uRu. Let u be an idempotent element of R, $uu = u$, and let uRu be the ring of all elements $x \in R$ with $ux = x = xu$ with the binary operation inherited from R and the neutral element u. This ring is Morita equivalent to R whenever u is *pseudoinvertible*; that is, $eum = 1$ for some elements e and m of R.

Morita's original result is that the two preceding operations are sufficient: if a ring S is Morita equivalent to R, that is, R-*Mod* and S-*Mod* are equivalent categories, then S is isomorphic to the ring $uR^{[k]}u$ for some pseudoinvertible idempotent $k \times k$ matrix u.

We now generalize Morita constructions to one-sorted algebraic theories and mention the S-sorted case later.

15.4 Definition Let (T, T) be a one-sorted algebraic theory.

1. The *matrix theory* $(T^{[k]}, T^{[k]})$ for $k = 1, 2, 3, \ldots$ is the one-sorted algebraic theory whose morphisms $f\colon p \to q$ are precisely the morphisms $f\colon kp \to kq$ of T; composition and identity morphisms are defined as in T, and $T^{[k]}\colon \mathcal{N} \to T^{[k]}$ takes the projection π_i^n to the morphism of $T(kn, k)$, which is the ith chosen projection of $kn = k \times \ldots \times k$ in T.
2. Let $u\colon 1 \to 1$ be an idempotent morphism of T ($u \cdot u = u$). We call u *pseudoinvertible* provided that there exist morphisms $m\colon 1 \to k$ and $e\colon k \to 1$ such that
$$e \cdot u^k \cdot m = \mathrm{id}_1.$$

The *idempotent modification* of $(\mathcal{T}, T)$ is the theory $(u\mathcal{T}u, uTu)$ whose morphisms $f: p \to q$ are precisely the morphisms of $\mathcal{T}$ satisfying $f \cdot u^p = f = u^q \cdot f$. The composition is defined as in $\mathcal{T}$, and the identity morphism on p is u^p. The functor $uTu: \mathcal{N} \to u\mathcal{T}u$ is the codomain restriction of T.

15.5 Remark

1. Both $\mathcal{T}^{[k]}$ and $u\mathcal{T}u$ are well defined. In fact, $\mathcal{T}^{[k]}$ has finite products with $p = 1 \times \ldots \times 1$: the ith projection is obtained from the ith projection in $\mathcal{T}$ of $kp = k \times \ldots \times k$. Also, $u\mathcal{T}u$ has finite products with $p = 1 \times \ldots \times 1$: the i-projection $\pi_i: p \to 1$ of $\mathcal{T}$ yields a morphism $u \cdot \pi_i: p \to 1$ of $u\mathcal{T}u$ ($i = 1, \ldots, k$), and these morphisms form a product $p = 1 \times \ldots \times 1$ in $u\mathcal{T}u$.
2. Observe that in the definition of Morita equivalence, the categories $Alg\,\mathcal{T}$ are treated as abstract categories: the forgetful functor does not play a role here.

15.6 Theorem *Let $(\mathcal{T}, T)$ be a one-sorted algebraic theory.*

1. *The matrix theories $\mathcal{T}^{[k]}$ are Morita equivalent to $\mathcal{T}$ for all $k > 0$*
2. *The idempotent modifications $u\mathcal{T}u$ are Morita equivalent to $\mathcal{T}$ for all pseudoinvertible idempotents u.*

Proof

1. Matrix theory $\mathcal{T}^{[k]}$. We have a full and faithful functor $\mathcal{T}^{[k]} \to \mathcal{T}$ defined on objects by $n \mapsto nk$ and on morphisms as the identity mapping. Every object of $\mathcal{T}$ is a retract of an object coming from $\mathcal{T}^{[k]}$: in fact, for every n, consider the diagonal morphism $\Delta: n \to nk = n \times \ldots \times n$. Consequently, $\mathcal{T}$ and $\mathcal{T}^{[k]}$ have the same idempotent completion. Thus, by 8.12, they are Morita equivalent.

2. For the idempotent modification $u\mathcal{T}u$, we consider $\mathcal{T}$ as a full subcategory of $(Alg\,\mathcal{T})^{op}$ via the Yoneda embedding (1.12)

$$Y_\mathcal{T}: \mathcal{T} \to (Alg\,\mathcal{T})^{op}.$$

Following 8.3, the idempotent $Y_\mathcal{T}(u): Y_\mathcal{T}(1) \to Y_\mathcal{T}(1)$ has a splitting in $(Alg\,\mathcal{T})^{op}$, say,

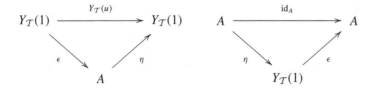

Consider also the subcategory $\mathcal{T}_A$ of $(Alg\,\mathcal{T})^{op}$ of all powers A^n, $n \in \mathbb{N}$. Together with the obvious functor $T_A \colon \mathcal{N} \to \mathcal{T}_A$, this is a one-sorted algebraic theory, and it is Morita equivalent to $\mathcal{T}$. In fact, every object of $\mathcal{T}$ is a retract of one in $\mathcal{T}_A$ and vice versa – this clearly implies that $\mathcal{T}$ and $\mathcal{T}_A$ have a joint idempotent completion (obtained by splitting their idempotents in $(Alg\,\mathcal{T})^{op}$). Indeed, since A is a retract of $Y_\mathcal{T}(1)$, A^p is a retract of $Y_\mathcal{T}(p)$. Conversely, consider $m\colon 1 \to n$ and $e\colon n \to 1$ in $\mathcal{T}$ such that $e \cdot u^n \cdot m = \mathrm{id}_1$, as in 15.4.2. Then $Y_\mathcal{T}(1)$ is a retract of A^n via $\epsilon^n \cdot Y_\mathcal{T}(m) \colon Y_\mathcal{T}(1) \to A^n$ and $Y_\mathcal{T}(e) \cdot \eta^n \colon A^n \to Y_\mathcal{T}(1)$, and $Y_\mathcal{T}(p)$ is a retract of A^{np}.

To complete the proof, we construct an equivalence functor $\bar{Y} \colon u\mathcal{T}u \to \mathcal{T}_A$. On objects, it is defined by $\bar{Y}(p) = A^p$ and on morphisms $f \colon p \to q$ by

$$\begin{array}{ccc} A^p & \xrightarrow{\bar{Y}f} & A^q \\ \eta^p \downarrow & & \uparrow \epsilon^q \\ Y_\mathcal{T}(p) & \xrightarrow[Y_\mathcal{T}(f)]{} & Y_\mathcal{T}(q) \end{array}$$

in $(Alg\,\mathcal{T})^{op}$. Observe that $\bar{Y}(\mathrm{id}_p) = \mathrm{id}_{A^p}$ because $\epsilon \cdot \eta = \mathrm{id}_A$. Now we check the equation

$$Y_\mathcal{T}(f) = \eta^q \cdot \bar{Y}(f) \cdot \epsilon^p. \tag{15.1}$$

Indeed,

$$Y_\mathcal{T}(f) = Y_\mathcal{T}(u)^q \cdot Y_\mathcal{T}(f) \cdot Y_\mathcal{T}(u)^p = \eta^q \cdot \epsilon^q \cdot Y_\mathcal{T}(f) \cdot \eta^p \cdot \epsilon^p = \eta^q \cdot \bar{Y}(f) \cdot \epsilon^p.$$

From Equation (15.1), since ϵ^p is a (split) epimorphism and η^q is is a (split) monomorphism, we deduce that $\bar{Y}$ preserves composition (because $Y_\mathcal{T}$ does) and that $\bar{Y}$ is faithful (because $Y_\mathcal{T}$ is). Since $\bar{Y}$ is surjective on objects, it remains to show that it is full: considering $h \colon A^p \to A^q$ in $(Alg\,\mathcal{T})^{op}$, we define $k = \eta^q \cdot h \cdot \epsilon^p \colon Y_\mathcal{T}(p) \to Y_\mathcal{T}(q)$. Since $Y_\mathcal{T}$ is full, there is a morphism $f \colon p \to q$ in $\mathcal{T}$ such that $Y_\mathcal{T}(f) = k$. Now

$$\bar{Y}(f) = \epsilon^q \cdot Y_\mathcal{T}(f) \cdot \eta^p = \epsilon^q \cdot \eta^q \cdot h \cdot \epsilon^p \cdot \eta^p = h.$$

It remains to check that f is in $u\mathcal{T}u$:

$$Y_\mathcal{T}(f) \cdot Y_\mathcal{T}(u^p) = \eta^q \cdot h \cdot \epsilon^p \cdot \eta^p \cdot \epsilon^p = \eta^q \cdot h \cdot \epsilon^p = k = Y_\mathcal{T}(f),$$

and then $f \cdot u^p = f$ because $Y_\mathcal{T}$ is faithful; analogously, $u^q \cdot f = f$. □

15.7 Theorem *For two one-sorted algebraic theories $(\mathcal{T}, T)$ and $(\mathcal{S}, S)$, the following conditions are equivalent:*

1. S is Morita equivalent to T
2. S is, as a category, equivalent to an idempotent modification $uT^{[k]}u$ of a matrix theory of T for some pseudoinvertible idempotent u of $T^{[k]}$

Proof Consider an equivalence functor
$$E\colon Alg\, S \to Alg\, T$$
and the Yoneda embeddings $Y_S\colon S^{op} \to Alg\, S$, $Y_T\colon T^{op} \to Alg\, T$ (recall, from 1.13, that Y_S and Y_T preserve finite coproducts). Since $Y_S(1)$ is perfectly presentable in $Alg\, S$ (see 5.5), we conclude that $A = E(Y_S(1))$ is perfectly presentable in $Alg\, T$ and therefore, due to 5.14, it is a retract of $Y_T(n)$ for some n in $\mathbb{N}$:

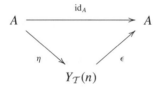

There exists a unique $u\colon n \to n$ in T such that $Y_T(u) = \eta \cdot \epsilon$, and such a u is an idempotent. We consider u as an idempotent on 1 in $T^{[n]}$ and prove that u is pseudoinvertible there. For this, choose an S-algebra $\bar{A}$ and an isomorphism $i\colon Y_T(n) \to E\bar{A}$. Since E is an equivalence functor, $\bar{A}$ is perfectly presentable, and thus it is a retract of $Y_S(k)$ for some $k \in \mathbb{N}$:

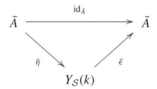

Consider now the composites

$$Y_T(n) \xrightarrow{i} E\bar{A} \xrightarrow{E\bar\eta} EY_S(k) \simeq kA \xrightarrow{k\eta} kY_T(n) \simeq Y_T(nk)$$
$$Y_T(nk) \simeq kY_T(n) \xrightarrow{k\epsilon} kA \simeq EY_S(k) \xrightarrow{E\bar\epsilon} E\bar{A} \xrightarrow{i^{-1}} Y_T(n)$$

Since Y_T is full, there exist unique morphisms $e\colon nk \to n$ and $m\colon n \to nk$ in T that Y_T maps on the preceding composites. One immediately checks that $Y_T(e \cdot u^k \cdot m) = \mathrm{id}$; that is, $e \cdot u^k \cdot m = \mathrm{id}$. Thus u is pseudoinvertible in $T^{[n]}$.

To complete the proof, we construct an equivalence functor $\bar{E}\colon S \to uT^{[n]}u$. It is the identity map on objects. If $f\colon p \to q$ is a morphism in S, $\bar{E}f$ is the unique morphism $np \to nq$ in $\mathcal{T}$ such that

$$\begin{array}{ccc}
qY_\mathcal{T}(n) \simeq Y_\mathcal{T}(nq) & \xrightarrow{Y_\mathcal{T}(\bar{E}f)} & Y_\mathcal{T}(np) \simeq pY_\mathcal{T}(n) \\
{\scriptstyle q\epsilon} \downarrow & & \uparrow {\scriptstyle p\eta} \\
qA \simeq E(Y_\mathcal{S}(q)) & \xrightarrow[E(Y_\mathcal{S}(f))]{} & E(Y_\mathcal{S}(p)) \simeq pA
\end{array}$$

commutes. Using, once again, $Y_\mathcal{T}(u) = \eta \cdot \epsilon$ and the faithfulness of $Y_\mathcal{T}$, one can easily check that $u^p \cdot \bar{E}f \cdot u^q = \bar{E}f$ so that $\bar{E}f$ is a morphism $p \to q$ in $u\mathcal{T}^{[n]}u$. The proof that $\bar{E}$ is a well-defined, full, and faithful functor is analogous to the proof in Theorem 15.6 and is left to the reader. $\square$

15.8 Example All one-sorted theories of *Set* are, up to equivalence of categories, precisely the theories $\mathcal{T}_k$ of 15.1. More precisely, for every k, consider the matrix theory $(\mathcal{N}^{[k]}, \mathrm{Id}^{[k]})$ (which, as a category, is clearly equivalent to $\mathcal{T}_k$ of 15.1). Given an idempotent $u: 1 \to 1$ of $\mathcal{N}^{[k]}$, the function $u: k \to k$ in *Set* is pseudoinvertible iff it is invertible; thus $u = \mathrm{id}$. Consequently, there are no other one-sorted theories of *Set*.

15.9 Example Let R be a ring with unit. Following 11.22, we can describe a one-sorted theory $(\mathcal{T}_R, T_R)$ of R-*Mod*: $\mathcal{T}_R$ is essentially the full subcategory of R-*Mod*^{op} of the finitely generated free R-modules R^n ($n \in \mathbb{N}$); that is, the morphisms in $\mathcal{T}_R(n, 1)$ are the homomorphisms from R to R^n, and T_R assigns to π_i^n the ith injection of $R + \ldots + R$. Every one-sorted algebraic theory of R-*Mod* is equivalent to $\mathcal{T}_S$ for some ring S that is Morita equivalent to R. Indeed, the two constructions of Example 15.3 fully correspond to the two constructions of Definition 15.4:

1. $\mathcal{T}_{(R^{[k]})}$ is equivalent to $(\mathcal{T}_R)^{[k]}$
2. given an idempotent element $u \in R$, the corresponding module homomorphism $\bar{u}: R \to R$ with $\bar{u}(x) = ux$ is such that uRu is equivalent to $\bar{u}(\mathcal{T}_R)\bar{u}$

15.10 Example For every monoid M, consider the category M-*Set* (cf. 13.15). Two monoids M and $\overline{M}$ are called *Morita equivalent* if M-*Set* and $\overline{M}$-*Set* are equivalent categories. Here we need just one operation on monoids: if $\overline{M}$ is Morita equivalent to M, then $\overline{M}$ is isomorphic to an idempotent modification uMu for some pseudoinvertible idempotent u of M.

In contrast with the situation of Example 15.9, M-*Set* has, in general, many one-sorted theories not connected to any Morita-equivalent monoid. (This is true even for $M = \{*\}$ since M-$Set = Set$ has infinitely many theories that are not equivalent as categories; see Example 15.1.) However, all *unary theories* of M-Set have a form that corresponds to Morita-equivalent monoids. By a unary theory, we mean a one-sorted theory $(\mathcal{T}, T)$ for which the category $\mathcal{T}$ is a free finite product completion (see 1.14) of the endomorphism monoid $\mathcal{T}(1, 1)$. The category of M-sets has an obvious unary theory with $\mathcal{T}(1, 1) = M$. Its morphisms from n to 1 are the homomorphisms from M to the free M-set $M + M + \ldots + M$ on n generators (so that the category $\mathcal{T}$ is, essentially, the full subcategory of $(M\text{-}Set)^{op}$ on the M-sets $M + M + \ldots + M$). Consequently, for every Morita-equivalent monoid $\overline{M}$, we have a unary theory $\mathcal{T}_{[\overline{M}]}$ for the category M-Set, and these are, up to categorical equivalence, all the unary theories. In fact, let $\mathcal{T}$ be a unary theory with $Alg\,\mathcal{T}$ equivalent to M-Set. For the monoid $\overline{M} = \mathcal{T}(T, T)$, there is an obvious categorical equivalence between $Alg\,\mathcal{T}$ and $\overline{M}$-Set: every $\overline{M}$-set $A: \overline{M} \to Set$ has an essentially unique extension to a $\mathcal{T}$-algebra $A': \mathcal{T} \to Set$, and $(-)'$ is the desired equivalence functor. Therefore $\overline{M}$ is Morita equivalent to M, and $\mathcal{T}$ is equivalent to $\mathcal{T}_{[\overline{M}]}$.

15.11 Remark

1. The preceding examples demonstrate that Theorems 15.6 and 15.7 yield a much more practical characterization than just stating that two theories have the same idempotent completion.
2. For S-sorted theories $(\mathcal{T}, T)$, the result is quite analogous. Given a collection

$$u = (u_s)_{s \in S}$$

with $u_s: s \to s$ idempotent, let us call u *pseudoinvertible* provided that for each $s \in S$, there exists a word $t_1 \ldots t_k$ and morphisms

$$m_s: s \to t_1 \ldots t_k \qquad e_s: t_1 \ldots t_k \to s$$

in $\mathcal{T}$ with

$$e_s \cdot (u_{t_1} \times \ldots \times u_{t_k}) \cdot m_s = \mathrm{id}_s.$$

Then $(u\mathcal{T}u, uTu)$ is the S-sorted theory whose morphisms $f: s_1 \ldots s_n \to r$ are those morphisms of $\mathcal{T}$ with $f \cdot (u_{s_1} \times \ldots \times u_{s_n}) = f = u_r \cdot f$, and the functor uTu is a codomain restriction of T.

For two S-sorted theories $(\mathcal{T}, T)$ and $(\mathcal{T}', T')$, we have that $\mathcal{T}'$ is Morita equivalent to $\mathcal{T}$ iff it is (as a category) equivalent to $u\mathcal{T}u$ for some pseudoinvertible collection $u = (u_s)_{s \in S}$. The proof is analogous to those of Theorems 15.6 and 15.7; the reader can find it in Adámek et al. (2006).

15.12 Remark Another approach to classical Morita theory for rings is based on the following result, due to Eilenberg and Watts (see Bass, 1968, Theorem 2.3): Let R, S be unitary rings, and let M be an R-S-bimodule. The formation of tensor products $M \otimes_S X$ for S-modules X defines a functor

$$M \otimes_S (-) : S\text{-Mod} \to R\text{-Mod}$$

that preserves colimits (because it is a left adjoint). In fact, the assignement

$$M \mapsto M \otimes_S (-)$$

induces a bijection between isomorphism classes of R-S-bimodules and isomorphism classes of colimit-preserving functors.

Using the Eilenberg–Watts theorem, one can prove that R and S are Morita equivalent iff there exist bimodules M, N and bimodule isomorphisms

$$M \otimes_S N \simeq R \qquad N \otimes_R M \simeq S.$$

These facts are easy to generalize to (abstract) algebraic theories. The generalization of the Eilenberg–Watts theorem in 15.17 uses the following lemma.

15.13 Lemma *Let $\mathcal{T}$ be an algebraic theory. The functor*

$$Y_\mathcal{T} : \mathcal{T}^{op} \to \mathrm{Alg}\,\mathcal{T}$$

is a free colimit completion of $\mathcal{T}^{op}$ conservative with respect to finite coproducts. This means that

1. *$\mathrm{Alg}\,\mathcal{T}$ is cocomplete and $Y_\mathcal{T}$ preserves finite coproducts*
2. *for every functor $F: \mathcal{T}^{op} \to \mathcal{B}$ preserving finite coproducts, where $\mathcal{B}$ is a cocomplete category, there exists an essentially unique functor F^*: $\mathrm{Alg}\,\mathcal{T} \to \mathcal{B}$ preserving colimits with F naturally isomorphic to $F^* \cdot Y_\mathcal{T}$*

Proof This follows from 1.13 and 4.15. □

15.14 Definition Let $\mathcal{T}, \mathcal{S}$ be algebraic theories. A *bimodule*

$$M: \mathcal{T} \Rightarrow \mathcal{S}$$

is a functor $M: \mathcal{T}^{op} \to \mathrm{Alg}\,\mathcal{S}$ preserving finite coproducts.

15.15 Remark

1. The functor $Y_\mathcal{T}: \mathcal{T}^{op} \to \mathrm{Alg}\,\mathcal{T}$ is a bimodule $\mathcal{T} \Rightarrow \mathcal{T}$. More generally, every morphism of theories $\mathcal{T} \to \mathcal{S}$ induces a bimodule $\mathcal{T} \Rightarrow \mathcal{S}$ by composition with $Y_\mathcal{S}$.

2. Every bimodule M has, by Lemma 15.13, an extension $M^*: Alg\,\mathcal{T} \to Alg\,\mathcal{S}$ preserving colimits. These are, up to natural isomorphism, the only colimit-preserving functors between algebraic categories.
3. Given bimodules $M: \mathcal{T} \Rightarrow \mathcal{S}$ and $N: \mathcal{S} \Rightarrow \mathcal{R}$, we define $N \circ M: \mathcal{T} \Rightarrow \mathcal{R}$ by $N^* \cdot M$. This composition is associative up to isomorphism and $Y_\mathcal{S} \circ M \simeq M \simeq M \circ Y_\mathcal{T}$. (In other words, the 2-category Th_{bim} we now define is in fact a *bicategory* in the sense of Bénabou (1967).)

15.16 Definition

1. The *2-category* Th_{bim} has
    objects: algebraic theories
    1-cells: bimodules
    2-cells: natural transformations
2. The *2-category* ALG_{colim} has
    objects: algebraic categories
    1-cells: colimit-preserving functors
    2-cells: natural transformations

15.17 Corollary

1. *The 2-categories Th_{bim} and ALG_{colim} are biequivalent. In fact, the assignment*

$$M: \mathcal{T} \Rightarrow \mathcal{S} \mapsto M^*: Alg\,\mathcal{T} \to Alg\,\mathcal{S}$$

of Remark 15.15 extends to a biequivalence $Th_{bim} \simeq ALG_{colim}$.
2. *Two algebraic theories $\mathcal{T}$ and $\mathcal{S}$ are Morita equivalent iff there exist bimodules $M: \mathcal{T} \Rightarrow \mathcal{S}$ and $N: \mathcal{S} \Rightarrow \mathcal{T}$ such that $N \circ M \simeq Y_\mathcal{T}$ and $M \circ N \simeq Y_\mathcal{S}$.*

In fact, $Th_{bim} \to ALG_{colim}$ is an equivalence on hom-categories by Lemma 15.13. The rest of the proof is obvious.

Historical remarks

The classical results concerning equivalences for categories of modules were proved by Morita (1958). Thirty years later, Dukarm (1988) proved a generalization to one-sorted algebraic theories.

For one-sorted theories, an approach to Morita equivalence via bimodules is due to Borceux and Vitale (1994). The many-sorted version of Morita equivalence in 15.11 is due to Adámek et al. (2006). Example 15.10 is due to Banaschewski (1972). For more on Morita equivalences of one-sorted theories see McKenzie (1996) and Porst (2000).

The Eilenberg–Watts theorem quoted in Remark 15.12 was independently proved by Eilenberg (1961) and Watts (1960). An exhaustive treatment of Morita theory for rings in terms of bimodules appears in the monograph of Bass (1968).

16
Free exact categories

We know that every algebraic category is an exact category having enough regular projective objects (see 3.18 and 5.15). In this chapter, we study *free exact completions* and prove that every algebraic category is a free exact completion of its full subcategory of all regular projectives. This will be used in the next chapter to characterize algebraic categories among exact categories and to describe all finitary localizations of algebraic categories.

The trouble with regular projective objects in an algebraic category is that they are not closed under finite limits. Luckily, they have weak finite limits. Recall that *weak limits* are defined as limits, except that the uniqueness of the factorization is not requested (see 16.7). The main point is that the universal property of a free exact completion is based on *left-covering functors*. These are functors that play, for categories with weak finite limits, the role that functor-preserving finite limits play for finitely complete categories.

We will be concerned with regular epimorphisms (3.4) in an exact category (3.16). For the comfort of the reader, we start by listing some of their (easy but) important properties. In diagrams, regular epimorphisms are denoted by $\longrightarrow\!\!\!\!\!\rightarrow$.

16.1 Lemma *Let $\mathcal{A}$ be an exact category.*

1. *Any morphism factorizes as a regular epimorphism followed by a monomorphism.*
2. *Consider a morphism $f\colon X \to Z$. The following conditions are equivalent:*
   a. *f is a regular epimorphism*
   b. *f is a strong epimorphism*
   c. *f is an extremal epimorphism*

Proof

1. Consider a morphism $f\colon X \to Z$ and its factorization through the coequalizer of its kernel pair

$$N(f) \underset{f_2}{\overset{f_1}{\rightrightarrows}} X \xrightarrow{f} Z$$

with $e\colon X \to I$ and $m\colon I \to Z$ such that $f = m \cdot e$.

We have to prove that m is a monomorphism. For this, consider the following diagram, where each square is a pullback:

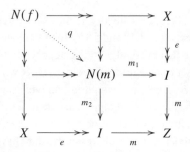

Since in $\mathcal{A}$, regular epimorphisms are pullback stable, the diagonal q is an epimorphism. Now $m_1 \cdot q = e \cdot f_1 = e \cdot f_2 = m_2 \cdot q$ so that $m_1 = m_2$. This means that m is a monomorphism.

2. For the implication a ⇒ b; let u, v, and m be morphisms such that $v \cdot f = m \cdot u$. If f is the coequalizer of a pair (x, y) and m a monomorphism, then u also coequalizes x and y.

For b ⇒ c; if $f = m \cdot u$ with m being a monomorphism, we can write $v \cdot f = m \cdot u$ with $v = \mathrm{id}$. By condition b, m is a split epimorphism, but a monomorphism that is also a split epimorphism is an isomorphism.

For c ⇒ a; just take a regular epi-mono factorization $f = m \cdot e$ (which exists by condition 1); if condition 16.1.c holds, then m is an isomorphism, and therefore f is a regular epimorphism. □

16.2 Corollary *Let $\mathcal{A}$ be an exact category.*

1. *The factorization stated in Lemma 16.1.1 is essentially unique.*
2. *The composite of two regular epimorphisms is a regular epimorphism.*

3. If the triangle

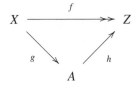

commutes and f is a regular epimorphism, then h is a regular epimorphism.
4. If a morphism is a regular epimorphism and a monomorphism, then it is an isomorphism.

In fact, everything follows easily from 16.1.2.

16.3 Lemma *Every exact category has the following properties:*

1. *The product of two regular epimorphisms is a regular epimorphism.*
2. *Consider the diagram*

$$\begin{array}{ccc} A_0 & \overset{a_1}{\underset{a_2}{\rightrightarrows}} & A_1 \\ {\scriptstyle f_0}\downarrow & & \downarrow{\scriptstyle f_1} \\ B_0 & \underset{b_2}{\overset{b_1}{\rightrightarrows}} & B_1 \end{array}$$

with $f_1 \cdot a_i = b_i \cdot f_0$ for $i = 1, 2$. If f_0 is a regular epimorphism and f_1 is a monomorphism, then the unique extension to the equalizers is a regular epimorphism.
3. *Consider the following commutative diagram:*

$$\begin{array}{ccccc} A_0 & \overset{a_1}{\longrightarrow} & A & \overset{a_2}{\longleftarrow} & A_1 \\ {\scriptstyle f_0}\downarrow & & {\scriptstyle f}\downarrow & & \downarrow{\scriptstyle f_1} \\ B_0 & \underset{b_1}{\longrightarrow} & B & \underset{b_2}{\longleftarrow} & B_1 \end{array}$$

If f_0 and f_1 are regular epimorphisms and f is a monomorphism, then the extension of f from the pullback of a_1 and a_2 to the pullback of b_1 and b_2 is a regular epimorphism.

Proof

1. Observe that $f \times \mathrm{id}$ is the pullback of f along the suitable projection and that the same holds for $\mathrm{id} \times g$. Now $f \times g = (f \times \mathrm{id}) \cdot (\mathrm{id} \times g)$.

2. Since f_1 is a monomorphism, the pullback of the equalizer of (b_1, b_2) along f_0 is the equalizer of (a_1, a_2).

3. This follows from points 1 and 2, using the usual construction of pullbacks via products and equalizers. □

For the sake of generality, let us point out that in 16.1, 16.2, and 16.3, we do not need that in $\mathcal{A}$ equivalence relations are effective.

16.4 Remark In 16.3.2, if f_1 is any morphism (not necessarily a monomorphism), it is no longer true that the pullback of an equalizer $e \colon E \to B_0$ of (b_1, b_2) along f_0 is an equalizer of (a_1, a_2). What remains true (in any category with finite limits) is the following fact: let $e' \colon E' \to A_0$ be a pullback of e along f_0, let $k_1, k_2 \colon N(f_1) \rightrightarrows A_1$ be a kernel pair of f_1, and let $n \colon E' \to N(f_1)$ be the unique morphism such that $k_i \cdot n = a_i \cdot e'$ ($i = 1, 2$). Then the following is a limit diagram:

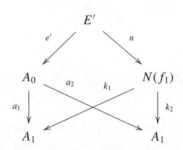

We will use this fact in the proof of Theorem 16.24.

From Propositions 3.18 and 5.15, we know that an algebraic category is an exact category having enough regular projective objects. In fact, each algebra is a regular quotient of a regular projective algebra. In the following, we study categories having enough regular projectives, and we introduce the concept of a regular projective cover for a subcategory of regular projectives in case there are "enough of them."

16.5 Definition Let $\mathcal{A}$ be a category. A *regular projective cover* of $\mathcal{A}$ is a full and faithful functor $I \colon \mathcal{P} \to \mathcal{A}$ such that

1. for every object P of $\mathcal{P}$, the object IP is regular projective in $\mathcal{A}$
2. for every object A of $\mathcal{A}$, there exists an object P in $\mathcal{P}$ and a regular epimorphism $P \to A$ (we write P instead of IP, and we call $P \to A$ a $\mathcal{P}$-cover of A)

16.6 Definition A functor is *exact* if it preserves finite limits and regular epimorphisms.

This chapter is devoted to the study of exact functors defined on an exact category $\mathcal{A}$ having a regular projective cover $\mathcal{P} \to \mathcal{A}$. First of all, observe that regular projective objects are not closed under finite limits so that we cannot hope that $\mathcal{P}$ inherits finite limits from $\mathcal{A}$. Nevertheless, a trace of finite limits remains in $\mathcal{P}$: In fact, $\mathcal{P}$ has weak finite limits.

16.7 Definition A *weak limit* of a diagram $D: \mathcal{D} \to \mathcal{A}$ is a cone $p_X: W \to DX$ ($X \in \mathrm{obj}\,\mathcal{D}$) such that for every other cone $a_X: A \to DX$, there exists a morphism $a: A \to W$ such that $p_X \cdot a = a_X$ for all X.

Observe that unlike limits, weak limits are very much nonunique. For example, any nonempty set is a weak terminal object in the category *Set*.

16.8 Lemma *If $\mathcal{P} \to \mathcal{A}$ is a regular projective cover of a finitely complete category $\mathcal{A}$, then $\mathcal{P}$ has weak finite limits.*

Proof Consider a finite diagram $D: \mathcal{D} \to \mathcal{P}$. If

$$\langle \pi_X: L \to DX \rangle_{X \in \mathcal{D}}$$

is a limit of D in $\mathcal{A}$, then we choose a $\mathcal{P}$-cover $l: P \to L$. The resulting cone

$$\langle \pi_X \cdot l: P \to DX \rangle_{X \in \mathcal{D}}$$

is a weak limit of D in $\mathcal{P}$. □

In the situation of the previous lemma, apply an exact functor $G: \mathcal{A} \to \mathcal{B}$. Since G preserves finite limits, the factorization of the cone

$$\langle G(\pi_X \cdot l): GP \to G(DX) \rangle_{X \in \mathcal{D}}$$

through the limit in $\mathcal{B}$ is $Gl: GP \to GL$, which is a regular epimorphism because G is exact. We can formalize this property in the following definition.

16.9 Definition Let $\mathcal{B}$ be an exact category and let $\mathcal{P}$ be a category with weak finite limits. A functor $F: \mathcal{P} \to \mathcal{B}$ is *left covering* if, for any finite diagram

$D: \mathcal{D} \to \mathcal{P}$ with weak limit W, the canonical comparison morphism $FW \to \lim F \cdot D$ is a regular epimorphism.

16.10 Remark To avoid any ambiguity in the previous definition, let us point out that if the comparison $w: FW \to \lim F \cdot D$ is a regular epimorphism for a certain weak limit W of D, then the comparison $w': FW' \to \lim F \cdot D$ is a regular epimorphism for any other weak limit W' of D. This follows from Corollary 16.2 because w factorizes through w'.

16.11 Example

1. If a finite diagram $D: \mathcal{D} \to \mathcal{A}$ has a limit L, then the weak limits of D are precisely the objects W such that L is a retract of W. Therefore any functor preserving finite limits is left covering.
2. If $\mathcal{P} \to \mathcal{A}$ is a regular projective cover of an exact category $\mathcal{A}$, then it is a left-covering functor.
3. The composition of a left-covering functor with an exact functor is a left-covering functor.

16.12 Example Let $\mathcal{P}$ be a category with weak finite limits, and consider the (possibly illegitimate) functor category $[\mathcal{P}^{op}, Set]$. The canonical Yoneda embedding $Y_{\mathcal{P}^{op}}: \mathcal{P} \to [\mathcal{P}^{op}, Set]$ is a left covering functor.

Proof Consider a finite diagram $D: \mathcal{D} \to \mathcal{P}$ in $\mathcal{P}$, a weak limit W of D, and a limit L of $Y_{\mathcal{P}^{op}} \cdot D$. The canonical comparison $\tau: Y_{\mathcal{P}^{op}}(W) \to L$ is a regular epimorphism whenever, for all $Z \in \mathcal{P}$, $\tau_Z: Y_{\mathcal{P}^{op}}(W)(Z) \to LZ$ is surjective. Since the limit L is computed pointwise in Set, an element of LZ is a cone from Z to $\mathcal{L}$ so that the surjectivity of τ_Z is just the weak universal property of W. □

16.13 Remark In the main result of this chapter (16.24), we show that an exact category with enough regular projective objects is a free exact completion of any of its regular projective covers. This is one of the results that requests working with the left-covering property (instead of the seemingly more natural condition of preservation of weak finite limits). In fact, the basic example 16.11.2 would not be true otherwise. This can be illustrated by the category of rings: the inclusion of the full subcategory $\mathcal{P}$ of all regular projective rings does not preserve weak finite limits. For example, the ring $\mathbb{Z}$ of integers is a weak terminal object in $\mathcal{P}$, but it is not a weak terminal object in $\mathcal{A}$ because the unique morphism from $\mathbb{Z}$ to the one-element ring does not have a section.

A remarkable fact about left-covering functors is that they classify exact functors. Before stating this in a precise way (see 16.24), we need some facts about left-covering functors and pseudoequivalences. A pseudoequivalence is defined almost as an equivalence relation but (1) using a weak pullback instead of a pullback to express the transitivity and (2) without the assumption that the graph be jointly monic.

16.14 Definition Let $\mathcal{P}$ be a category with weak pullbacks. A *pseudoequivalence* is a parallel pair

$$X' \underset{x_2}{\overset{x_1}{\rightrightarrows}} X$$

which is

1. reflexive, that is, there exists $r\colon X \to X'$ such that $x_1 \cdot r = \mathrm{id}_X = x_2 \cdot r$
2. symmetric, that is, there exists $s\colon X' \to X'$ such that $x_1 \cdot s = x_2$ and $x_2 \cdot s = x_1$
3. transitive, that is, in an arbitrary weak pullback,

$$\begin{array}{ccc} P & \xrightarrow{x'_1} & X' \\ {\scriptstyle x'_2}\downarrow & & \downarrow{\scriptstyle x_2} \\ X' & \xrightarrow{x_1} & X \end{array}$$

there exists $t\colon P \to X'$ such that $x_1 \cdot t = x_1 \cdot x'_1$ and $x_2 \cdot t = x_2 \cdot x'_2$. The morphism t is called a *transitivity morphism* of x_1 and x_2.

16.15 Remark

1. Observe that the existence of a transitivity morphism of x_1 and x_2 does not depend on the choice of a weak pullback of x_1 and x_2.
2. Recall that a regular factorization of a morphism is a factorization as a regular epimorphism followed by a monomorphism. In a category with binary products, we speak about *regular factorization of a parallel pair* $p, q\colon A \rightrightarrows B$. What we mean is a factorization of (p, q) as in the following diagram, where e is a regular epimorphism and (p', q') is a jointly

monomorphic parallel pair,

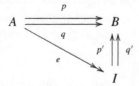

obtained by a regular factorization of $\langle p, q \rangle : A \to B \times B$. Since jointly monomorphic parallel pairs are also called *relations*, we call (p', q') the *relation induced* by (p, q).

3. If $\mathcal{P}$ has finite limits, then equivalence relations are precisely those parallel pairs that are, at the same time, relations and pseudoequivalences. The next result, which is the main link between pseudoequivalences and left-covering functors, shows that any pseudoequivalence in an exact category is a composition of an equivalence relation with a regular epimorphism. (The converse is not true: if we compose an equivalence relation with a regular epimorphism, in general, we do not obtain a pseudoequivalence. Consider the category of rings, the unique equivalence relation on the one-element ring 0, and the unique morphism $\mathbb{Z} \to 0$. The parallel pair $\mathbb{Z} \rightrightarrows 0$ is not reflexive because there are no morphisms from 0 to $\mathbb{Z}$.)

16.16 Lemma *Let $F: \mathcal{P} \to \mathcal{B}$ be a left-covering functor. For every pseudoequivalence $x_1, x_2: X' \rightrightarrows X$ in $\mathcal{P}$, the relation in $\mathcal{B}$ induced by (Fx_1, Fx_2) is an equivalence relation.*

Proof Consider a regular factorization in $\mathcal{B}$:

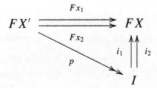

Since the reflexivity and transitivity are obvious, we only check the transitivity of (i_1, i_2). The pullback of $i_1 \cdot p$ and $i_2 \cdot p$ factorizes through the pullback of i_1 and i_2, and the factorization, v, is a regular epimorphism (because p is a regular epimorphism and $\mathcal{B}$ is an exact category):

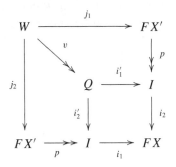

Consider also a transitivity morphism $t: P \to X'$ of (x_1, x_2), as in Definition 16.14. Since $F: \mathcal{P} \to \mathcal{B}$ is left covering, the factorization $q: FP \to W$ such that $j_1 \cdot q = Fx'_1$ and $j_2 \cdot q = Fx'_2$ is a regular epimorphism. Finally, we have the following commutative diagram:

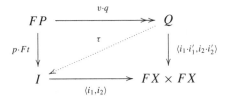

Since $v \cdot q$ is a regular epimorphism and $\langle i_1, i_2 \rangle$ is a monomorphism, there exists $\tau: Q \to I$ such that $\langle i_1, i_2 \rangle \cdot \tau = \langle i_1 \cdot i'_1, i_2 \cdot i'_2 \rangle$. This implies that (i_1, i_2) is transitive. □

16.17 Remark Generalizing the fact that functors preserve finite limits iff they preserve finite products and equalizers, we are going to prove the same for left covering functors. We use the phrase "left covering with respect to weak finite products" for the restriction of 16.9 to discrete categories $\mathcal{D}$. Observe that this is equivalent to being left covering with respect to weak binary products and weak terminal objects. Analogously, we use "left covering with respect to weak equalizers."

16.18 Lemma *A functor $F: \mathcal{P} \to \mathcal{B}$, where $\mathcal{P}$ has weak finite limits and $\mathcal{B}$ is exact, is left covering iff it is left covering with respect to weak finite products and weak equalizers.*

Proof
1. Using Lemma 16.3 and working by induction, one extends the left covering character of F to joint equalizers of parallel n-tuples and then to multiple pullbacks.

2. Consider a finite diagram $D\colon \mathcal{D} \to \mathcal{P}$. We can construct a weak limit of D using a weak product $\Pi_{X\in\mathcal{D}} DX$, weak equalizers E_d, one for each morphism $d\colon X \to X'$ in $\mathcal{D}$, and a weak multiple pullback E, as in the following diagram:

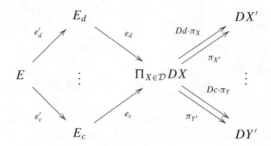

Perform the same constructions in $\mathcal{B}$ to get limits, as in the following diagrams:

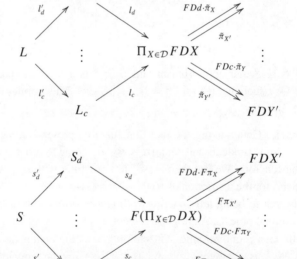

By assumption, the canonical factorization $q_d\colon FE_d \to S_d$ is a regular epimorphism. By Lemma 16.3, this gives rise to a regular epimorphism $q\colon Q \to S$, where Q is the multiple pullback of the Fe_d. By part 1, the canonical factorization $t\colon FE \to Q$ is a regular epimorphism. Finally, a diagram chase shows that the pullback of $l_d \cdot l'_d$ along the canonical factorization $p: F(\Pi_{X \in \mathcal{D}} DX) \to \Pi_{X \in \mathcal{D}} FDX$ is $s_d \cdot s'_d$. By part 1, p is a regular epimorphism so that we get a regular epimorphism $p': S \to L$. The regular epimorphism $p' \cdot q \cdot t: FE \to Q \to S \to L$ shows that F is left covering. □

16.19 Lemma *A left-covering functor $F: \mathcal{P} \to \mathcal{B}$ preserves finite jointly monomorphic sources.*

Proof A family of morphisms $(f_i: A \to A_i)_{i \in I}$ is jointly monomorphic iff the span formed by id_A, id_A is a limit of the corresponding diagram.

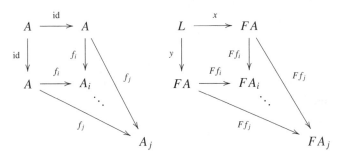

Now apply F and consider the canonical factorization $q: FA \to L$, where L is a limit in $\mathcal{B}$ of the corresponding diagram with the limit cone $x, y: L \to FA$. By assumption, q is a regular epimorphism. It is also a monomorphism because $x \cdot q = \text{id}$, and so it is an isomorphism. This implies that $\text{id}_{FA}, \text{id}_{FA}$ is a limit, thus the family $(Ff_i: FA \to FA_i)_{i \in I}$ is jointly monomorphic. □

16.20 Lemma *Consider a functor $F: \mathcal{P} \to \mathcal{B}$. Assume that $\mathcal{P}$ has finite limits and $\mathcal{B}$ is exact. Then F is left covering iff it preserves finite limits.*

Proof One implication is clear (see 16.11.1). Thus let us assume that F is left covering, and consider a finite nonempty diagram $D: \mathcal{D} \to \mathcal{P}$. Let $(\pi_X: L \to DX)_{X \in \mathcal{D}}$ be a limit of D and $(\tilde{\pi}_X: \tilde{L} \to FDX)_{X \in \mathcal{D}}$ a limit of $F \cdot D$. Since the family $(\pi_X)_{X \in \mathcal{D}}$ is jointly monomorphic, by Lemma 16.19 the family $(F\pi_X)_{X \in \mathcal{D}}$ is also monomorphic. This implies that the canonical factorization $p: FL \to \tilde{L}$ is a monomorphism. But it is a regular epimorphism by assumption, so that it is an isomorphism.

The argument for the terminal object T is different. In $\mathcal{P}$, the product of T with itself is T, with the identity morphisms as projections. Then the canonical factorization $FT \to FT \times FT$ is a (regular) epimorphism. This implies that the two projections $\pi_1, \pi_2: FT \times FT \rightrightarrows FT$ are equal. But the pair (π_1, π_2) is the kernel pair of the unique morphism q to the terminal object of $\mathcal{B}$ so that

q is a monomorphism. Since F is left covering, q is a regular epimorphism and thus an isomorphism. □

Let us point out that in 16.16 and 16.18, we do not need to assume that equivalence relations are effective in $\mathcal{B}$. Moreover, if in Definition 16.9 we replace *regular epimorphism* by *strong epimorphism*, then 16.19 and 16.20 hold for all categories $\mathcal{B}$ with finite limits.

16.21 Definition Let $\mathcal{P}$ be a category with weak finite limits. A *free exact completion* of $\mathcal{P}$ is an exact category $\mathcal{P}_{ex}$ with a left-covering functor

$$\Gamma \colon \mathcal{P} \to \mathcal{P}_{ex}$$

such that for every exact category $\mathcal{B}$ and for every left-covering functor $F \colon \mathcal{P} \to \mathcal{B}$, there exists an essentially unique exact functor $\hat{F} \colon \mathcal{P}_{ex} \to \mathcal{B}$ with $\hat{F} \cdot \Gamma$ naturally isomorphic to F.

Note that since a free exact completion is defined via a universal property, it is determined uniquely up to equivalence.

16.22 Remark Since the composition of the left-covering functor $\Gamma \colon \mathcal{P} \to \mathcal{P}_{ex}$ with an exact functor $\mathcal{P}_{ex} \to \mathcal{B}$ clearly gives a left covering functor $\mathcal{P} \to \mathcal{B}$, the previous universal property can be restated in the following way: composition with Γ induces an equivalence functor

$$- \cdot \Gamma \colon Ex[\mathcal{P}_{ex}, \mathcal{B}] \to Lco[\mathcal{P}, \mathcal{B}],$$

where $Ex[\mathcal{P}_{ex}, \mathcal{B}]$ is the category of exact functors from $\mathcal{P}_{ex}$ to $\mathcal{B}$ and natural transformations, and $Lco[\mathcal{P}, \mathcal{B}]$ is the category of left-covering functors from $\mathcal{P}$ to $\mathcal{B}$ and natural transformations.

16.23 Remark To prepare the proof of Theorem 16.24, let us explain how an exact category with enough regular projective objects can be reconstructed using any of its regular projective covers. Let $\mathcal{P} \to \mathcal{A}$ be a regular projective cover of an exact category $\mathcal{A}$. Fix an object A in $\mathcal{A}$ and consider a $\mathcal{P}$-cover $a \colon X \to A$, its kernel pair $a_1, a_2 \colon N(a) \rightrightarrows X$, and, again, a $\mathcal{P}$-cover $x \colon X' \to N(a)$. In the resulting diagram,

$$X' \underset{a_2 \cdot x}{\overset{a_1 \cdot x}{\rightrightarrows}} X \overset{a}{\twoheadrightarrow} A,$$

the left-hand part is a pseudoequivalence in $\mathcal{P}$ (not in $\mathcal{A}$!), and A is its coequalizer. Consider a morphism $\varphi \colon A \to B$ in $\mathcal{A}$ and the following diagram:

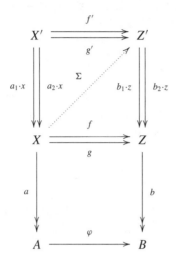

Using the regular projectivity of X and X' and the universal property of the kernel pair of b, we get a pair (f', f) such that $\varphi \cdot a = b \cdot f$ and $f \cdot a_i \cdot x = b_i \cdot z \cdot f'$ for $i = 1, 2$. Conversely, a pair (f', f) such that $f \cdot a_i \cdot x = b_i \cdot z \cdot f'$ for $i = 1, 2$ induces a unique extension to the quotient. Moreover, two such pairs (f', f) and (g', g) have the same extension iff there is a morphism $\Sigma \colon X \to Z'$ such that $b_1 \cdot z \cdot \Sigma = f$ and $b_2 \cdot z \cdot \Sigma = g$.

16.24 Theorem *Let $I \colon \mathcal{P} \to \mathcal{A}$ be a regular projective cover of an exact category $\mathcal{A}$. Then $\mathcal{A}$ is a free exact completion of $\mathcal{P}$.*

Proof
1. For extension of a left-covering functor $F \colon \mathcal{P} \to \mathcal{B}$ to a functor $\hat{F} \colon \mathcal{A} \to \mathcal{B}$, define $\hat{F}$ on objects $A \in \mathcal{A}$ by constructing the coequalizer:

$$X' \underset{x_2 = a_2 \cdot x}{\overset{x_1 = a_1 \cdot x}{\rightrightarrows}} X \overset{a}{\twoheadrightarrow} A$$

as in 16.23. By 16.16, the relation (i_1, i_2) induced by $Fx_1, Fx_2 \colon FX' \rightrightarrows FX$ in $\mathcal{B}$ is an equivalence relation. Since $\mathcal{B}$ is exact, we can define $\hat{F}A$ to be a coequalizer of (i_1, i_2):

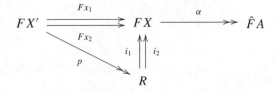

Let now $\varphi: A \to B$ be a morphism in $\mathcal{A}$. We can construct a pair $f: X \to Z$, $f': X' \to Z'$ as in 16.23 and define $\hat{F}\varphi$ to be the unique extension to the quotients as in the following diagram:

$$\begin{array}{ccc} FX' & \xrightarrow{Ff'} & FZ' \\ {\scriptstyle Fx_1} \downdownarrows {\scriptstyle Fx_2} & & {\scriptstyle Fz_1} \downdownarrows {\scriptstyle Fz_2} \\ FX & \xrightarrow{Ff} & FZ \\ {\scriptstyle \alpha} \downarrow & & \downarrow {\scriptstyle \beta} \\ \hat{F}A & \xrightarrow{\hat{F}\varphi} & \hat{F}B \end{array}$$

The discussion in 16.23 shows that this definition does not depend on the choice of the pair f, f'. The preservation of composition and identity morphisms by $\hat{F}$ comes from the uniqueness of the extension to the quotients. It is clear that $\hat{F} \cdot I$ is naturally isomorphic to F and that a different choice of $\mathcal{P}$-covers X and X' for a given object $A \in \mathcal{A}$ produces a functor naturally isomorphic to $\hat{F}$.

2. $\hat{F}$ is the essentially unique exact functor such that $\hat{F} \cdot I$ is naturally isomorphic to F. Indeed, using once again the notations of 16.23, $(\hat{F}a_1, \hat{F}a_2)$ is the kernel pair of $\hat{F}a$, and $\hat{F}x$ and $\hat{F}a$ are regular epimorphisms:

$$FX' \simeq \hat{F}X' \xrightarrow{\hat{F}x} \hat{F}N(a) \underset{\hat{F}a_2}{\overset{\hat{F}a_1}{\rightrightarrows}} \hat{F}X \simeq FX \xrightarrow{\hat{F}a} \hat{F}A.$$

This implies that $\hat{F}A$ is necessarily a coequalizer of (Fx_1, Fx_2). In a similar way, one shows that $\hat{F}$ is uniquely determined on morphisms.

3. The extension $\hat{F}: \mathcal{A} \to \mathcal{B}$ preserves finite limits. In fact, it is sufficient to show that $\hat{F}$ is left covering with respect to the terminal object, binary products, and equalizers of pairs (see Lemmas 16.18 and 16.20).

3a. For products; let A and B be objects in $\mathcal{A}$. Working as in 16.23, we get coequalizers

$$X' \underset{x_2}{\overset{x_1}{\rightrightarrows}} X \overset{a}{\twoheadrightarrow} A \qquad Z' \underset{z_2}{\overset{z_1}{\rightrightarrows}} Z \overset{b}{\twoheadrightarrow} B.$$

Consider the following diagram, where both horizontal lines are products in $\mathcal{A}$ and $c: R \to X \times Z$ is a $\mathcal{P}$-cover:

$$\begin{array}{ccccc}
 & & R & & \\
 & & \downarrow c & & \\
X & \overset{\pi_X}{\leftarrow} & X \times Z & \overset{\pi_Z}{\to} & Z \\
a \downarrow & & \downarrow a \times b & & \downarrow b \\
A & \underset{\pi_A}{\leftarrow} & A \times B & \underset{\pi_B}{\to} & B
\end{array}$$

By 16.3.1, $a \times b$ is a regular epimorphism so that $(a \times b) \cdot c: R \to A \times B$ is a $\mathcal{P}$-cover. Moreover, by 16.8, $(R, \pi_X \cdot c, \pi_Z \cdot c)$ is a weak product of X and Z in $\mathcal{P}$. Applying $\hat{F}$, we have the following diagram in $\mathcal{B}$:

$$\begin{array}{ccccc}
FX & \overset{F(\pi_X \cdot c)}{\leftarrow} & FR & \overset{F(\pi_Z \cdot c)}{\to} & FZ \\
\alpha \downarrow & & \gamma \downarrow & & \downarrow \beta \\
\hat{F}A & \underset{\hat{F}\pi_A}{\leftarrow} & \hat{F}(A \times B) & \underset{\hat{F}\pi_B}{\to} & \hat{F}B
\end{array}$$

from which we get the following commutative square:

$$\begin{array}{ccc}
FR & \overset{\langle F(\pi_X \cdot c), F(\pi_Z \cdot c) \rangle}{\twoheadrightarrow} & FX \times FZ \\
\gamma \downarrow & & \downarrow \alpha \times \beta \\
\hat{F}(A \times B) & \underset{\langle \hat{F}\pi_A, \hat{F}\pi_B \rangle}{\to} & \hat{F}A \times \hat{F}B
\end{array}$$

The top morphism is a regular epimorphism because F is left covering, and the right-hand morphism is a regular epimorphism by 16.3.3 so that the bottom morphism also is a regular epimorphism, as requested.

3b. For equalizers; let

$$E \xrightarrow{e} A \underset{\psi}{\overset{\varphi}{\rightrightarrows}} B$$

be an equalizer in $\mathcal{A}$. The idea is once again to construct a $\mathcal{P}$-cover of E using $\mathcal{P}$-covers of A and B. For that, consider the following diagram:

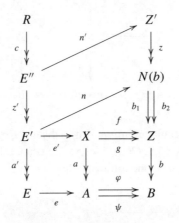

where $a: X \to A$ and $b: Z \to B$ are $\mathcal{P}$-covers, E' is a pullback of a and e, the equations $b \cdot f = \varphi \cdot a$ and $b \cdot g = \psi \cdot a$ hold, $N(b)$ is a kernel pair of b, n is the unique morphism such that $b_1 \cdot n = f \cdot e'$ and $b_2 \cdot n = g \cdot e'$, E'' is a pullback of z and n, and $z: Z' \to N(b)$ and $c: R \to E''$ are $\mathcal{P}$-covers. From 16.4, we know that the following is a limit diagram in $\mathcal{A}$:

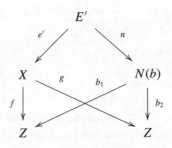

Clearly it remains a limit diagram if we paste it with the pullback E'', and by 16.8, we get a weak limit in $\mathcal{P}$ by covering it with $c\colon R \to E''$:

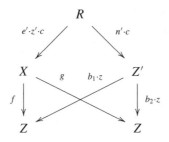

Consider the following diagram in $\mathcal{B}$:

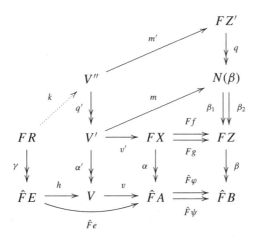

where V is an equalizer of $\hat{F}\varphi$ and $\hat{F}\psi$; $\alpha\colon FX \to \hat{F}A$, $\beta\colon FZ \to \hat{F}B$, and $\gamma\colon FR \to \hat{F}E$ are the coequalizers defining $\hat{F}A$, $\hat{F}B$, and $\hat{F}E$ (as explained in the first part of the proof), V' is a pullback of α and v, $N(\beta)$ is a kernel pair of β, m is the unique morphism such that $\beta_1 \cdot m = Ff \cdot v'$ and $\beta_2 \cdot m = Fg \cdot v'$, q is the unique morphism such that $\beta_i \cdot q = F(b_i \cdot z)$ $(i = 1, 2)$, V'' is a pullback of q and m; and $h\colon \hat{F}E \to V$ is the unique morphism such that $v \cdot h = \hat{F}e$. We have to prove that h is a regular epimorphism. By 16.4, the following is a limit diagram:

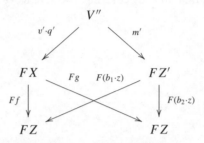

Since F is left covering, the unique morphism $k\colon FR \to V''$ such that $m' \cdot k = F(n' \cdot c)$ and $v' \cdot q' \cdot k = F(e' \cdot z' \cdot c)$ is a regular epimorphism. As $\mathcal{B}$ is exact and F is left covering, by 16.16, (β_1, β_2) is the equivalence relation induced by (z_1, z_2), so that q is a regular epimorphism, and then q' also is a regular epimorphism. Since α' also is a regular epimorphism, it remains to check that $\alpha' \cdot q' \cdot k = h \cdot \gamma$. By composing with the monomorphism v, this is an easy diagram chasing.

3c. For a terminal object; let $1_{\mathcal{A}}$ and $1_{\mathcal{B}}$ be terminal objects of $\mathcal{A}$ and $\mathcal{B}$ and $T \to 1_{\mathcal{A}}$ a $\mathcal{P}$-cover. Applying $\hat{F}$, we get a commutative diagram

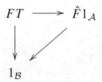

Since T is weak terminal in $\mathcal{P}$ and F is left covering, $FT \to 1_{\mathcal{B}}$ is a regular epimorphism. Therefore $\hat{F}1_{\mathcal{A}} \to 1_{\mathcal{B}}$ also is a regular epimorphism.

4. The extension $\hat{F}\colon \mathcal{A} \to \mathcal{B}$ preserves regular epimorphisms. This is obvious: if $\varphi\colon A \to B$ is a regular epimorphism in $\mathcal{A}$ and $a\colon X \to A$ is a $\mathcal{P}$-cover, we can choose as a $\mathcal{P}$-cover of B the morphism $\varphi \cdot a\colon X \to B$. Applying $\hat{F}$, we get a commutative diagram

$$\begin{array}{ccc} FX & \xrightarrow{\mathrm{id}} & FX \\ {\scriptstyle \alpha}\downarrow & & \downarrow{\scriptstyle \beta} \\ \hat{F}A & \xrightarrow[\hat{F}\varphi]{} & \hat{F}B \end{array}$$

which shows that $\hat{F}\varphi$ is a regular epimorphism. □

16.25 Corollary

1. Let $\mathcal{A}$ and $\mathcal{B}$ be exact categories and $I: \mathcal{P} \to \mathcal{A}$ a regular projective cover. Given exact functors $G, G': \mathcal{A} \rightrightarrows \mathcal{B}$ with $G \cdot I \simeq G' \cdot I$, then $G \simeq G'$.
2. Let $\mathcal{A}$ and $\mathcal{A}'$ be exact categories, $\mathcal{P} \to \mathcal{A}$ a regular projective cover of $\mathcal{A}$, and $\mathcal{P}' \to \mathcal{A}'$ one of $\mathcal{A}'$. Any equivalence $\mathcal{P} \simeq \mathcal{P}'$ extends to an equivalence $\mathcal{A} \simeq \mathcal{A}'$.

16.26 Remark For later use, let us point out a simple consequence of the previous theorem. Consider the free exact completion $I: \mathcal{P} \to \mathcal{A}$, as in 16.24, and a functor $K: \mathcal{A} \to \mathcal{B}$, with $\mathcal{B}$ exact. If K preserves coequalizers of equivalence relations and $K \cdot I$ is left covering, then K is exact.

16.27 Corollary Let $\mathcal{A}$ be an algebraic category and $\mathcal{P}$ its full subcategory of regular projective objects. The inclusion $I: \mathcal{P} \to \mathcal{A}$ is a free exact completion of $\mathcal{P}$.

In fact, this follows from Theorem 16.24 because $\mathcal{A}$ is exact (see 3.18) and $\mathcal{P}$ is a projective cover of $\mathcal{A}$ (see 5.15).

Historical remarks

Following a suggestion of A. Joyal, the exact completion of a category with finite limits was presented by Carboni and Celia Magno (1982). The more general approach working with categories with weak finite limits is from Carboni and Vitale (1998). The connection between the exact completion and the homotopy category of topological spaces (see 17.4) was established by Gran and Vitale (1998).

17
Exact completion and reflexive-coequalizer completion

This chapter is devoted to elementary constructions of two free completions described in Chapters 16 and 7, respectively, in a different manner: a free exact completion (of categories with weak finite limits) and a free reflexive-coequalizer completion (of categories with finite coproducts). The reader may decide to skip this chapter without losing the connection to Chapter 18.

In Chapter 16, we have seen that if $I: \mathcal{P} \to \mathcal{A}$ is a regular projective cover of an exact category $\mathcal{A}$, then $\mathcal{P}$ has weak finite limits, and I is a free exact completion of $\mathcal{P}$. We complete here the study of the exact completion by showing that for any category $\mathcal{P}$ with weak finite limits, it is possible to construct a free exact completion $\Gamma: \mathcal{P} \to \mathcal{P}_{ex}$. Moreover, Γ is a regular projective cover of $\mathcal{P}_{ex}$.

The following construction of $\mathcal{P}_{ex}$ is suggested by 16.23.

17.1 Definition Given a category $\mathcal{P}$ with weak finite limits, we define the category $\mathcal{P}_{ex}$ as follows:

1. Objects of $\mathcal{P}_{ex}$ are pseudoequivalences $x_1, x_2: X' \rightrightarrows X$ in $\mathcal{P}$ (we sometimes denote such an object by X/X').
2. A *premorphism* in $\mathcal{P}_{ex}$ is a pair of morphisms (f', f) as in the diagram

$$\begin{array}{ccc} X' & \xrightarrow{f'} & Z' \\ {\scriptstyle x_1} \downarrow\downarrow {\scriptstyle x_2} & & {\scriptstyle z_1} \downarrow\downarrow {\scriptstyle z_2} \\ X & \xrightarrow{f} & Z \end{array}$$

such that $f \cdot x_1 = z_1 \cdot f'$ and $f \cdot x_2 = z_2 \cdot f'$.

3. A *morphism* in $\mathcal{P}_{ex}$ is an equivalence class $[f', f]: X/X' \to Z/Z'$ of premorphisms, where two parallel premorphisms (f', f) and (g', g) are equivalent if there exists a morphism $\Sigma: X \to Z'$ such that $z_1 \cdot \Sigma = f$ and $z_2 \cdot \Sigma = g$.
4. Composition and identities are obvious.

17.2 Notation We denote by $\Gamma: \mathcal{P} \to \mathcal{P}_{ex}$ the embedding of $\mathcal{P}$ into $\mathcal{P}_{ex}$, assigning to a morphism $f: X \to Z$ the following morphism:

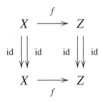

17.3 Remark

1. The fact that the preceding relation among premorphisms is an equivalence relation can be proved (step by step) using the assumption that the codomain $z_1, z_2: Z' \rightrightarrows Z$ is a pseudoequivalence. Observe also that the class of (f', f) depends on f only (compose f with a reflexivity morphism of (z_1, z_2) to show that (f', f) and (f'', f) are equivalent); for this reason, we often write $[f]$ instead of $[f', f]$.
2. The fact that composition is well defined is obvious.
3. Γ is a full and faithful functor. This is easy to verify.
4. Observe that if $\mathcal{P}$ is small (or locally small), then $\mathcal{P}_{ex}$ also is small (or locally small, respectively).

17.4 Remark The preceding equivalence relation among premorphisms in $\mathcal{P}_{ex}$ can be thought of as a kind of homotopy relation. And in fact, this is the case in a particular example: let X be a topological space and $X^{[0,1]}$ the space of continuous maps from the interval $[0, 1]$ to X; the evaluation maps $ev_0, ev_1: X^{[0,1]} \rightrightarrows X$ constitute a pseudoequivalence. This gives rise to a functor $\mathcal{E}: \textbf{Top} \to \textbf{Top}_{ex}$. Now two continuous maps $f, g: X \to Z$ are homotopic in the usual sense precisely when $\mathcal{E}(f)$ and $\mathcal{E}(g)$ are equivalent in the sense of Definition 17.1. More precisely, $\mathcal{E}$ factorizes through the homotopy category, and the factorization $\mathcal{E}': \textbf{HTop} \to \textbf{Top}_{ex}$ is full and faithful (and left covering).

We are going to prove that the preceding category $\mathcal{P}_{ex}$ is exact and the functor $\Gamma: \mathcal{P} \to \mathcal{P}_{ex}$ is a regular projective cover. For this, it is useful to have an equivalent description of $\mathcal{P}_{ex}$ as a full subcategory of the functor category $[\mathcal{P}^{op}, \textbf{Set}]$.

17.5 Lemma *Let $\mathcal{P}$ be a category with weak finite limits, and let*

$$Y_{\mathcal{P}^{op}} \colon \mathcal{P} \to [\mathcal{P}^{op}, Set]$$

be the Yoneda embedding. The following properties of a functor $A \colon \mathcal{P}^{op} \to Set$ are equivalent:

1. *A is a regular quotient of a representable object modulo a pseudoequivalence in $\mathcal{P}$; that is, there exists a pseudoequivalence $x_1, x_2 \colon X' \rightrightarrows X$ in $\mathcal{P}$ and a coequalizer*

$$Y_{\mathcal{P}^{op}}(X') \underset{x_2}{\overset{x_1}{\rightrightarrows}} Y_{\mathcal{P}^{op}}(X) \twoheadrightarrow A$$

   in $[\mathcal{P}^{op}, Set]$

2. *A is a regular quotient of a representable object modulo a regular epimorphism $a \colon Y_{\mathcal{P}^{op}}(X) \to A$ such that $N(a)$, the domain of a kernel pair of a, is also a regular quotient of a representable object:*

$$Y_{\mathcal{P}^{op}}(X') \overset{x}{\twoheadrightarrow} N(a) \underset{a_2}{\overset{a_1}{\rightrightarrows}} Y_{\mathcal{P}^{op}}(X) \overset{a}{\twoheadrightarrow} A \quad \text{(for some } X' \text{ in } \mathcal{P}\text{)}$$

Proof Consider the morphisms of point 2. Since a is the coequalizer of $(a_1 \cdot x, a_2 \cdot x)$, we have to prove that $(a_1 \cdot x, a_2 \cdot x)$ is a pseudoequivalence in $\mathcal{P}$. Let us check the transitivity: consider the following diagram:

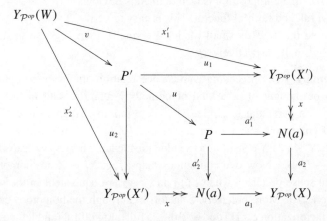

where P and P' are pullbacks and W is a weak pullback. Since $Y_{\mathcal{P}^{op}}(W)$ is regular projective and x is a regular epimorphism, the transitivity morphism $t \colon P \to N(a)$ of (a_1, a_2) extends to a morphism $t' \colon Y_{\mathcal{P}^{op}}(W) \to Y_{\mathcal{P}^{op}}(X')$ such that $t \cdot u \cdot v = x \cdot t'$. This morphism t' is a transitivity morphism for $(a_1 \cdot x, a_2 \cdot x)$. The converse implication follows from Lemma 16.16 since $Y_{\mathcal{P}^{op}} \colon \mathcal{P} \to [\mathcal{P}^{op}, Set]$ is left covering (see 16.12).

Note that (x_1, x_2) is a pseudoequivalence in $\mathcal{P}$ does not mean that $(Y_{\mathcal{P}^{op}}(x_1), Y_{\mathcal{P}^{op}}(x_2))$ is a pseudoequivalence in $[\mathcal{P}^{op}, Set]$ because $Y_{\mathcal{P}^{op}}$ does not preserve weak pullbacks. □

17.6 Remark The full subcategory of $[\mathcal{P}^{op}, Set]$ of all objects satisfying 1 or 2 of 17.5 is denoted by $\mathcal{P}'_{ex}$. In the next lemma, the codomain restriction of the Yoneda embedding $Y_{\mathcal{P}^{op}}: \mathcal{P} \to [\mathcal{P}^{op}, Set]$ to $\mathcal{P}'_{ex}$ is again denoted by $Y_{\mathcal{P}^{op}}$, and Γ is the functor from 17.2.

17.7 Lemma *There exists an equivalence of categories* $\mathcal{E}: \mathcal{P}_{ex} \to \mathcal{P}'_{ex}$ *such that* $\mathcal{E} \cdot \Gamma = Y_{\mathcal{P}^{op}}$.

Proof Consider the functor $\mathcal{E}: \mathcal{P}_{ex} \to \mathcal{P}'_{ex}$ sending a morphism $[f]: X/X' \to Z/Z'$ to the corresponding morphism φ between the coequalizers, as in the following diagram:

$$\begin{array}{ccc} Y_{\mathcal{P}^{op}}(X') & \xrightarrow{f'} & Y_{\mathcal{P}^{op}}(Z') \\ {\scriptstyle x_1}\downarrow\downarrow{\scriptstyle x_2} & & {\scriptstyle z_1}\downarrow\downarrow{\scriptstyle z_2} \\ Y_{\mathcal{P}^{op}}(X) & \xrightarrow{f} & Y_{\mathcal{P}^{op}}(Z) \\ {\scriptstyle a}\downarrow & & \downarrow{\scriptstyle b} \\ A & \xrightarrow{\varphi} & B \end{array}$$

The functor $\mathcal{E}$ is well defined because a is an epimorphism and b coequalizes y_0 and y_1. Moreover, $\mathcal{E}$ is essentially surjective by definition of $\mathcal{P}'_{ex}$. Let us prove that $\mathcal{E}$ is faithful: if $\mathcal{E}[f] = \mathcal{E}[g]$, then the pair (f, g) factorizes through the kernel pair $N(b)$ of b, which is a regular factorization of (y_0, y_1). Since $Y_{\mathcal{P}^{op}}(X)$ is regular projective, this factorization extends to a morphism $Y_{\mathcal{P}^{op}}(X) \to Y_{\mathcal{P}^{op}}(Z')$, which shows that $[f] = [g]$.

The functor $\mathcal{E}$ is full: given $\varphi: A \to B$, we get $f: Y_{\mathcal{P}^{op}}(X) \to Y_{\mathcal{P}^{op}}(Z)$ by regular projectivity of $Y_{\mathcal{P}^{op}}(X)$. Since $b \cdot f \cdot x_1 = b \cdot f \cdot x_2$, we get $\overline{f}: Y_{\mathcal{P}^{op}}(X') \to N(b)$. Since $N(b)$ is the regular factorization of (z_1, z_2) and $Y_{\mathcal{P}^{op}}(X')$ is regular projective, $\overline{f}$ extends to $f': Y_{\mathcal{P}^{op}}(X') \to Y_{\mathcal{P}^{op}}(Z')$. Clearly $\mathcal{E}[f', f] = \varphi$. □

17.8 Proposition *For every category $\mathcal{P}$ with weak finite limits, the functor*

$$\Gamma: \mathcal{P} \to \mathcal{P}_{ex}$$

of 17.2 is a left-covering functor into an exact category. Moreover, this is a regular projective cover of $\mathcal{P}_{ex}$.

Proof

1. $\mathcal{P}_{ex}$ has finite limits. Since the construction of the other basic types of finite limits is completely analogous, we explain in detail the case of equalizers, mentioning construction for binary products and terminal objects just briefly.

1a. For equalizers; consider a parallel pair in $\mathcal{P}_{ex}$ together with what we want to be their equalizer:

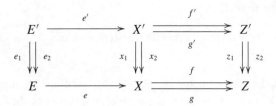

This means that we need the following equations: $x_1 \cdot e' = e \cdot e_1$ and $x_2 \cdot e' = e \cdot e_2$. Moreover, we request $f \cdot e$ and $g \cdot e$ to be equivalent; that is, we need a morphism $\varphi \colon E \to Z'$ such that $z_1 \cdot \varphi = f \cdot e$ and $z_2 \cdot \varphi = g \cdot e$. Let us take E and E' to be the following weak limits:

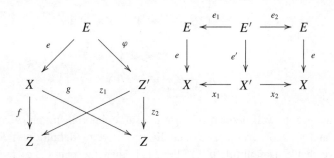

i. It is straightforward to check that (e_1, e_2) is a pseudoequivalence in $\mathcal{P}$ (just use the fact that (x_1, x_2) is a pseudoequivalence).

ii. To show that $[e]$ equalizes $[f]$ and $[g]$, use the morphism $\varphi \colon E \to Z'$.

iii. The morphism $[e]$ is a monomorphism: in fact, consider two morphisms in $\mathcal{P}_{ex}$,

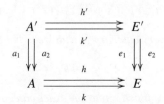

Exact completion and reflexive-coequalizer completion   187

such that $[e] \cdot [h] = [e] \cdot [k]$. This means that there is a morphism $\Sigma: A \to X'$ such that $x_1 \cdot \Sigma = e \cdot h$ and $x_2 \cdot \Sigma = e \cdot k$. By the weak universal property of E', we have a morphism $\Sigma': A \to E'$ such that $e_1 \cdot \Sigma' = h$ and $e_2 \cdot \Sigma' = k$. This means that $[h] = [k]$.

iv. We prove that every morphism

$$
\begin{array}{ccc}
A' & \xrightarrow{h'} & X' \\
{\scriptstyle a_1} \downarrow\downarrow {\scriptstyle a_2} & & {\scriptstyle x_1} \downarrow\downarrow {\scriptstyle x_2} \\
A & \xrightarrow{h} & X
\end{array}
$$

in $\mathcal{P}_{ex}$ such that $[f] \cdot [h] = [g] \cdot [h]$ factorizes through $[e]$. We know that there is $\Sigma: A \to Z'$ such that $z_1 \cdot \Sigma = f \cdot h$ and $z_2 \cdot \Sigma = g \cdot h$. The weak universal property of E yields, then, a morphism $k: A \to E$ such that $e \cdot k = h$ and $\Sigma \cdot k = \varphi$. Now $x_1 \cdot h' = e \cdot k \cdot a_1$ and $x_2 \cdot h' = e \cdot k \cdot a_2$. The weak universal property of E' yields a morphism $k': A' \to E'$ such that $e_1 \cdot k' = k \cdot a_1$ and $e_2 \cdot k' = k \cdot a_2$. Finally, the needed factorization is $[k', k]: A/A' \to E/E'$.

1b. For products; consider two objects $x_1, x_2: X' \rightrightarrows X$ and $z_1, z_2: Z' \rightrightarrows Z$ in $\mathcal{P}_{ex}$. Their product is given by

$$
\begin{array}{ccccc}
X' & \xleftarrow{x'} & P' & \xrightarrow{z'} & Z' \\
{\scriptstyle x_1} \downarrow\downarrow {\scriptstyle x_2} & {\scriptstyle p_1} & \downarrow\downarrow & {\scriptstyle p_2} & {\scriptstyle z_1} \downarrow\downarrow {\scriptstyle z_2} \\
X & \xleftarrow{x} & P & \xrightarrow{z} & Z
\end{array}
$$

where

$$
X \xleftarrow{x} P \xrightarrow{z} Z
$$

is a weak product of X and Z in $\mathcal{P}$, and P' is the following weak limit:

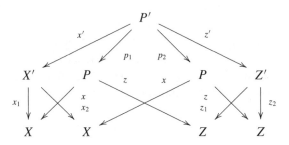

1c. For terminal objects; consider an object T of $\mathcal{P}$, the projections from a weak product $\pi_1, \pi_2 \colon T \times T \rightrightarrows T$ form a pseudoequivalence. If T is a weak terminal object in $\mathcal{P}$, then (π_1, π_2) is a terminal object in $\mathcal{P}_{ex}$.

2. $\mathcal{P}_{ex}$ is closed under finite limits in $[\mathcal{P}^{op}, Set]$. In fact, by Lemma 17.7, we can identify $\mathcal{P}_{ex}$ with $\mathcal{P}'_{ex}$. We prove that the full inclusion of $\mathcal{P}_{ex}$ into $[\mathcal{P}^{op}, Set]$ preserves finite limits. Because of Lemma 16.20, it is enough to prove that the inclusion is left covering. We give the argument for equalizers since that for products and terminal objects is similar (and easier). With the notations of part 1, consider the following diagram, where ϵ, α, and β are extensions to the coequalizers, the triangle on the right is a regular factorization, and the triangle at the bottom is the factorization through the equalizer:

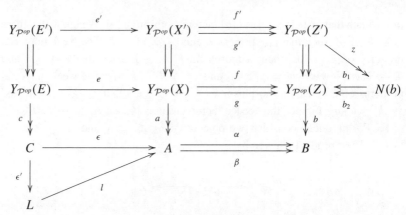

We have to prove that ϵ' is a regular epimorphism. Using $\varphi \colon E \to Z'$, we check that $\alpha \cdot a \cdot e = \beta \cdot a \cdot e$ so that there is $p \colon Y_{\mathcal{P}^{op}}(E) \to L$ such that $l \cdot p = a \cdot e$, and then $p = \epsilon' \cdot c$. So it is enough to prove that p is a regular epimorphism; that is, the components $p(P) \colon Y_{\mathcal{P}^{op}}(E)(P) \to L(P)$ are surjective. This means that given a morphism $u \colon Y_{\mathcal{P}^{op}}(P) \to A$ such that $\alpha \cdot u = \beta \cdot u$, we need a morphism $\hat{u} \colon P \to E$ with $l \cdot p \cdot \hat{u} = u$. First of all, observe that since a is a regular epimorphism and $Y_{\mathcal{P}^{op}}(P)$ is regular projective, there is $u' \colon P \to X$ such that $a \cdot u' = u$. Now $b \cdot f \cdot u' = b \cdot g \cdot u'$ so that there is $u'' \colon Y_{\mathcal{P}^{op}}(P) \to N(b)$ such that $b_1 \cdot u'' = f \cdot u'$ and $b_2 \cdot u'' = g \cdot u'$. Moreover, since z is a regular epimorphism and $Y_{\mathcal{P}^{op}}(P)$ is regular projective, there is $\tilde{u} \colon P \to Z'$ with $z \cdot \tilde{u} = u''$. Finally, $z_1 \cdot \tilde{u} = f \cdot u'$ and $z_2 \cdot \tilde{u} = g \cdot u'$ so that there is $\hat{u} \colon P \to E$ such that $\varphi \cdot \hat{u} = \tilde{u}$ and $e \cdot \hat{u} = u'$. This last equation implies that $l \cdot p \cdot \hat{u} = u$.

3. $\mathcal{P}_{ex}$ is closed in $[\mathcal{P}^{op}, Set]$ under coequalizers of equivalence relations. In fact, consider an equivalence relation in $\mathcal{P}_{ex}$, with its coequalizer in $[\mathcal{P}^{op}, Set]$:

$$B \underset{\beta}{\overset{\alpha}{\rightrightarrows}} A \overset{c}{\twoheadrightarrow} C.$$

We have to prove that C lies in $\mathcal{P}_{ex}$. For this, consider the following diagram:

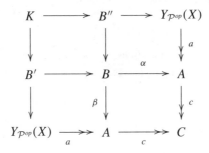

with each square, except possibly the right-hand bottom one, being a pullback. The remaining square is, then, also a pullback because $[\mathcal{P}^{op}, Set]$ is exact, $X \in \mathcal{P}$, and a is a regular epimorphism. Since A, B and $Y_{\mathcal{P}^{op}}(X)$ are in $\mathcal{P}_{ex}$, which is closed in $[\mathcal{P}^{op}, Set]$ under finite limits (see part 2), K lies in $\mathcal{P}_{ex}$ too. So K is a regular quotient of a representable object. But K is also the kernel pair of the regular epimorphism $c \cdot a: Y_{\mathcal{P}^{op}}(X) \to C$. By Lemma 17.5, this means that C is in $\mathcal{P}_{ex}$. □

17.9 Corollary *For every category $\mathcal{P}$ with weak finite limits, the functor*

$$\Gamma: \mathcal{P} \to \mathcal{P}_{ex}$$

of 17.2 is a free exact completion of $\mathcal{P}$.

In fact, this follows from 16.24 and 17.8.

17.10 Remark Let $\mathcal{A}$ be an algebraic category. From Chapter 7, we know that there are equivalences

$$\mathcal{A} \simeq Ind(\mathcal{A}_{fp}) \quad \text{and} \quad \mathcal{A}_{fp} \simeq Rec(\mathcal{A}_{pp}),$$

where $\mathcal{A}_{pp}$ and $\mathcal{A}_{fp}$ are the full subcategories of $\mathcal{A}$ of perfectly presentable objects and of finitely presentable objects, respectively. Recall from 17.11 that Rec is the free completion under finite colimits conservative with respect to finite coproducts. An analogous situation holds with the exact completion. In fact, there are equivalences

$$\mathcal{A} \simeq (\mathcal{A}_{rp})_{ex} \simeq (FCSum(\mathcal{A}_{pp}))_{ex} \quad \text{and} \quad \mathcal{A}_{rp} \simeq Ic(FCSum(\mathcal{A}_{pp})),$$

where $\mathcal{A}_{rp}$ is the full subcategory of regular projective objects and $FCSum$ is the free completion under coproducts conservative with respect to finite coproducts.

The second part of this chapter is devoted to an elementary description of the free completion under reflexive coequalizers

$$E_{Rec}: \mathcal{C} \to Rec\,\mathcal{C}$$

of a category $\mathcal{C}$ with finite coproducts, already studied in Chapter 7. Let us start by observing that 7.3 can be restated as follows.

17.11 Proposition *Let $\mathcal{T}$ be an algebraic theory. The functor*

$$Y_{\mathcal{T}}: \mathcal{T}^{op} \to (Alg\,\mathcal{T})_{fp}$$

is a free completion of $\mathcal{T}$ under finite colimits, conservative with respect to finite coproducts. This means that

1. *$(Alg\,\mathcal{T})_{fp}$ is finitely cocomplete and $Y_{\mathcal{T}}$ preserves finite coproducts*
2. *for every functor $F: \mathcal{T}^{op} \to \mathcal{B}$ preserving finite coproducts, where $\mathcal{B}$ is a finitely cocomplete category, there exists an essentially unique functor $F^*: Alg\,\mathcal{T} \to \mathcal{B}$ preserving finite colimits with F naturally isomorphic to $F^* \cdot Y_{\mathcal{T}}$*

We pass now to an elementary description of $Rec\,\mathcal{C}$.

17.12 Definition Given a category $\mathcal{C}$ with finite coproducts, we define the category $Rec\,\mathcal{C}$ as follows:

1. Objects of $Rec\,\mathcal{C}$ are reflexive pairs $x_1, x_2: X_1 \rightrightarrows X_0$ in $\mathcal{C}$ (i.e., parallel pairs for which there exists $d: X_0 \to X_1$ such that $x_1 \cdot d = \mathrm{id}_{X_0} = x_2 \cdot d$, see 3.12).
2. Consider the following diagram in $\mathcal{C}$,

$$V \underset{g}{\overset{f}{\rightrightarrows}} Z_0 \underset{z_2}{\overset{z_1}{\leftleftarrows}} Z_1$$

with z_1, z_2, being a reflexive pair. We write

$$h: f \mapsto g$$

if there exists a morphism $h: V \to Z_1$ such that $z_1 \cdot h = f$ and $z_2 \cdot h = g$. This is a reflexive relation in the hom-set $\mathcal{C}(V, Z_0)$. We write $f \sim g$ if f and g are in the equivalence relation generated by this reflexive relation.

Exact completion and reflexive-coequalizer completion 191

3. A *premorphism* in $Rec\,\mathcal{C}$ from (x_1, x_2) to (z_1, z_2) is a morphism f in $\mathcal{C}$ as in the diagram

$$\begin{array}{ccc} X_1 & & Z_1 \\ x_1 \downdownarrows x_2 & & z_1 \downdownarrows z_2 \\ X_0 & \xrightarrow{f} & Z_0 \end{array}$$

such that $f \cdot x_1 \sim f \cdot x_2$.
4. A *morphism* in $Rec\,\mathcal{C}$ from (x_1, x_2) to (z_1, z_2) is an equivalence class $[f]$ of premorphisms with respect to the equivalence $\sim$ of 2.
5. Composition and identities in $Rec\,\mathcal{C}$ are the obvious ones.
6. The functor $E_{Rec}: \mathcal{C} \to Rec\,\mathcal{C}$ is defined by

$$E_{Rec}(X \xrightarrow{f} Z) \quad = \quad \begin{array}{ccc} X & & Z \\ id \downdownarrows id & & id \downdownarrows id \\ X & \xrightarrow{[f]} & Z \end{array}$$

17.13 Remark

1. Consider

$$\begin{array}{c} Z_1 \\ z_1 \downdownarrows z_2 \\ V \underset{g}{\overset{f}{\rightrightarrows}} Z_0 \end{array}$$

as in 17.12. Explicitly, $f \sim g$ means that there exists a zigzag

2. Using the explicit description of $f \sim g$, it is straightforward to prove that $Rec\,\mathcal{C}$ is a category and $E_{Rec}: \mathcal{C} \to Rec\,\mathcal{C}$ is a full and faithful functor.
3. The preceding description of $E_{Rec}: \mathcal{C} \to Rec\,\mathcal{C}$ does not depend on the existence of finite coproducts in $\mathcal{C}$.

17.14 Lemma *Let $\mathcal{C}$ be a category with finite coproducts. The category $\operatorname{Rec}\mathcal{C}$ of 17.12 has finite colimits, and $E_{Rec}\colon \mathcal{C} \to \operatorname{Rec}\mathcal{C}$ preserves finite coproducts.*

Proof
1. Finite coproducts in $\operatorname{Rec}\mathcal{C}$ are computed componentwise; that is, if x_1, x_2: $X_1 \rightrightarrows X_0$ and $z_1, z_2 \colon Z_1 \rightrightarrows Z_0$ are objects of $\operatorname{Rec}\mathcal{C}$, their coproduct is

$$\begin{array}{ccccc}
X_1 & & X_1 + Z_1 & & Z_1 \\
x_1 \Big\Downarrow x_2 & & x_1+z_1 \Big\Downarrow x_2+z_2 & & z_1 \Big\Downarrow z_2 \\
X_0 & \xrightarrow{[i_{X_0}]} & X_0 + Z_0 & \xleftarrow{[i_{Z_0}]} & Z_0
\end{array}$$

2. Reflexive coequalizers in $\operatorname{Rec}\mathcal{C}$ are depicted in the following diagram:

$$\begin{array}{ccccc}
X_1 & & Z_1 & & X_0 + Z_1 \\
x_1 \Big\Downarrow x_2 & & z_1 \Big\Downarrow z_2 & & \langle f,z_1 \rangle \Big\Downarrow \langle g,z_2 \rangle \\
X_0 & \underset{[g]}{\overset{[f]}{\rightrightarrows}} & Z_0 & \xrightarrow{[\operatorname{id}]} & Z_0
\end{array}$$

$\square$

17.15 Lemma *Consider the diagram*

$$\begin{array}{c}
Z_1 \\
z_1 \Big\Downarrow z_2 \\
V \underset{g}{\overset{f}{\rightrightarrows}} Z_0
\end{array}$$

as in 17.12. If a morphism $w\colon Z_0 \to W$ is such that $w \cdot z_1 = w \cdot z_2$ and $f \sim g$, then $w \cdot f = w \cdot g$.

Proof Clearly, if $h\colon f \mapsto g$, then $w \cdot f = w \cdot g$. The claim now follows from the fact that to be coequalized by w is an equivalence relation in $\mathcal{C}(V, Z_0)$. $\square$

17.16 Remark For every reflexive pair $x_1, x_2\colon X_1 \rightrightarrows X_0$ in $\mathcal{C}$, the diagram

$$E_{Rec} X_1 \underset{E_{Rec} x_2}{\overset{E_{Rec} x_1}{\rightrightarrows}} E_{Rec} X_0 \xrightarrow{[\operatorname{id}_{X_0}]} (X_1 \underset{x_2}{\overset{x_1}{\rightrightarrows}} X_0)$$

is a reflexive coequalizer in $\mathcal{R}ec\,\mathcal{C}$. Therefore, if two functors $F, G\colon \mathcal{R}ec\,\mathcal{C} \to \mathcal{B}$ preserve reflexive coequalizers and $F \cdot E_{Rec} \simeq G \cdot E_{Rec}$, then $F \simeq G$.

17.17 Proposition *Let $\mathcal{C}$ be a category with finite coproducts. The functor*

$$E_{Rec}\colon \mathcal{C} \to \mathcal{R}ec\,\mathcal{C}$$

of 17.12 is a free completion of $\mathcal{C}$ under finite colimits, conservative with respect to finite coproducts. This means that

1. *$\mathcal{R}ec\,\mathcal{C}$ has finite colimits and E_{Rec} preserves finite coproducts*
2. *for every functor $F\colon \mathcal{C} \to \mathcal{B}$ preserving finite coproducts, where $\mathcal{B}$ is a finitely cocomplete category, there exists an essentially unique functor $F^*\colon \mathcal{R}ec\,\mathcal{C} \to \mathcal{B}$ preserving finite colimits with F naturally isomorphic to $F^* \cdot E_{Rec}$:*

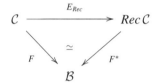

Proof Given $F\colon \mathcal{C} \to \mathcal{B}$, as before, we define $F^*\colon \mathcal{R}ec\,\mathcal{C} \to \mathcal{B}$ on objects by the following coequalizer in $\mathcal{B}$:

$$FX_1 \underset{Fx_2}{\overset{Fx_1}{\rightrightarrows}} FX_0 \longrightarrow F^*(x_1, x_2).$$

Lemma 17.15 makes it clear how to define F^* on morphisms. The argument for the essential uniqueness of F^* is stated in 17.16. The rest of the proof is straightforward. □

The previous universal property allows us to give a different proof of the equation $\mathcal{S}ind\,\mathcal{C} \simeq \mathcal{I}nd\,(\mathcal{R}ec\,\mathcal{C})$ already established in 7.4.

17.18 Corollary *Let $\mathcal{C}$ be a small category with finite coproducts. There exists an equivalence of categories*

$$\mathcal{I}nd\,(\mathcal{R}ec\,\mathcal{C}) \simeq \mathcal{S}ind\,\mathcal{C}.$$

We have proved this fact in 7.4. Here we obtain a different proof based on 17.17: Let $\mathcal{B}$ be a cocomplete category. By 4.18, the functors $\mathcal{I}nd\,(\mathcal{R}ec\,\mathcal{C}) \to \mathcal{B}$ preserving colimits correspond to the functors $\mathcal{R}ec\,\mathcal{C} \to \mathcal{B}$ preserving finite colimits and then, by 17.17, to the functors $\mathcal{C} \to \mathcal{B}$ preserving finite coproducts. On the other hand, the functors $\mathcal{C} \to \mathcal{B}$ preserving finite coproducts correspond, by 15.13, to the functors $Sind\,\mathcal{C} \to \mathcal{B}$ preserving colimits. Since both $Ind(Rec\,\mathcal{C})$ and $Sind\,\mathcal{C}$ are cocomplete (4.16 and 4.5), we can conclude that $Ind(Rec\,\mathcal{C})$ and $Sind\,\mathcal{C}$ are equivalent categories. □

Historical remarks

The reflexive coequalizer completion of a category with finite coproducts is due to A. M. Pitts (unpublished notes, 1996). It appeared in press in Bunge and Carboni (1995). The connection between the exact completion and the reflexive coequalizer completion was established by Pedicchio and Rosický (1999); see also Rosický and Vitale (2001) for the connection with homological functors.

18
Finitary localizations of algebraic categories

We know from Chapter 6 that algebraic categories are precisely the cocomplete categories having a strong generator formed by perfectly presentable objects. We prove now that among exact categories, the algebraic categories are precisely the cocomplete categories having a strong generator formed by finitely presentable regular projectives. As a consequence, we fully characterize all finitary localizations of algebraic categories as the exact, locally finitely presentable categories.

18.1 Theorem *Let $\mathcal{E}$ be an exact category with sifted colimits, $\mathcal{A}$ a category with finite limits and sifted colimits, and $F: \mathcal{E} \to \mathcal{A}$ a functor preserving finite limits and filtered colimits. Then F preserves reflexive coequalizers iff it preserves regular epimorphisms.*

Proof Necessity is evident because by 3.4 every regular epimorphism is a reflexive coequalizer of its kernel pair. For sufficiency, let F preserve finite limits, filtered colimits, and regular epimorphisms.

1. Since every equivalence relation in $\mathcal{E}$ is a kernel pair of its coequalizer and since every regular epimorphism is a coequalizer of its kernel pair, F preserves coequalizers of equivalence relations. Since every pseudoequivalence in $\mathcal{E}$ can be decomposed as a regular epimorphism followed by an equivalence relation (cf. 16.16), F preserves coequalizers of pseudoequivalences.

2. Consider a reflexive and symmetric pair $r = (r_1, r_2: X' \rightrightarrows X)$ of morphisms in $\mathcal{E}$. We construct a pseudoequivalence $\bar{r}$ containing r (the *transitive hull* of r) as a (filtered) colimit of the chain of compositions:

$$r \circ r \circ \ldots \circ r \quad n\text{-times};$$

the composition $r \circ r$ is depicted in the following diagram, where the square is a pullback:

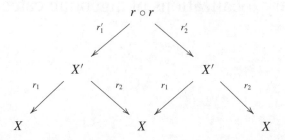

Since F preserves filtered colimits and finite limits, we have $\overline{Fr} = F\overline{r}$. The pseudoequivalence $\overline{r}$ has a coequalizer, which is preserved by F. But a coequalizer of $\overline{r}$ is also a coequalizer of r, and so F preserves coequalizers of reflexive and symmetric pairs of morphisms.

3. If $r = (r_1, r_2)$ is just a reflexive pair, then a reflexive and symmetric pair containing r is given by $r \circ r^{-1}$; that is,

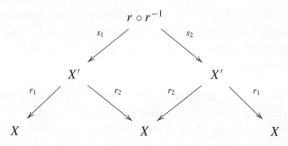

Once again, a coequalizer of $r \circ r^{-1}$ is also a coequalizer of r so that F preserves reflexive coequalizers. □

18.2 Corollary *Let $\mathcal{E}$ be a cocomplete exact category, $\mathcal{A}$ a category with finite limits and sifted colimits, and $F: \mathcal{E} \to \mathcal{A}$ a functor preserving finite limits. Then F preserves sifted colimits iff it preserves filtered colimits and regular epimorphisms.*

In fact, this follows from 7.7 and 18.1.

We can now generalize 5.16.

18.3 Corollary *In a cocomplete exact category, perfectly presentable objects are precisely finitely presentable regular projectives.*

Proof One implication is established in 5.4. For the converse implication, apply 18.2 to the hom-functor $\hom(G, -)$ of a finitely presentable regular projective object G. □

18.4 Corollary *A category is algebraic iff it is cocomplete, exact, and has a strong generator consisting of finitely presentable regular projectives.*

Proof Necessity follows from 3.18 and 6.9. Sufficiency follows from 18.3 and 6.9. □

In the previous corollary, the assumption of cocompleteness can be reduced to asking for the existence of coequalizers of kernel pairs, which is part of the exactness of the category, and the existence of coproducts of objects from the generator. In fact, we have the following result (recall that a category is *well powered* if, for a fixed object A, the subobjects of A constitute a set, not a proper class).

18.5 Lemma *Let $\mathcal{A}$ be a well-powered exact category with a regular projective cover $\mathcal{P} \to \mathcal{A}$. If $\mathcal{P}$ has coproducts, then $\mathcal{A}$ is cocomplete.*

Proof
1. The functor $\mathcal{P} \to \mathcal{A}$ preserves coproducts. Indeed, consider a coproduct

$$s_i \colon P_i \to \coprod_I P_i$$

in $\mathcal{P}$ and a family of morphisms $\langle x_i \colon P_i \to X \rangle_I$ in $\mathcal{A}$. Let $q \colon Q \to X$ be a regular epimorphism with $Q \in \mathcal{P}$. For each $i \in I$, consider a morphism $y_i \colon P_i \to Q$ such that $q \cdot y_i = x_i$. Since Q is in $\mathcal{P}$, there is $y \colon \coprod_I P_i \to Q$ such that $y \cdot s_i = y_i$, and then $q \cdot y \cdot s_i = x_i$, for all $i \in I$.

As far as the uniqueness of the factorization is concerned, consider a pair of morphisms $f, g \colon \coprod_I P_i \rightrightarrows X$ such that $s_i \cdot f = s_i \cdot g$ for all i. Consider also $f', g' \colon \coprod_I P_i \rightrightarrows Q$ such that $q \cdot f' = f$ and $q \cdot g' = g$. Since $q \cdot f' \cdot s_i = q \cdot g' \cdot s_i$, there is $t_i \colon P_i \to N(q)$ such that $q_1 \cdot t_i = f' \cdot s_i$ and $q_2 \cdot t_i = g' \cdot s_i$, where $q_1, q_2 \colon N(q) \rightrightarrows Q$ is a kernel pair of q. From the first part of the proof, we obtain a morphism $t \colon \coprod_I P_i \to N(q)$ such that $t \cdot s_i = t_i$ for all i. Moreover, $q_1 \cdot t \cdot s_i = f' \cdot s_i$ for all i so that $q_1 \cdot t = f'$ because Q is in $\mathcal{P}$. Analogously, $q_2 \cdot t = g'$. Finally, $f = q \cdot f' = q \cdot q_1 \cdot t = q \cdot q_2 \cdot t = q \cdot g' = g$.

2. Denote by $Sub_{\mathcal{A}}(A)$ the poset of subobjects of A. For every category $\mathcal{A}$, we denote by $\theta(\mathcal{A})$ its *ordered reflection*, that is, the ordered class obtained from the preorder on the objects of $\mathcal{A}$ given by $A \leq B$ iff $\mathcal{A}(A, B)$ is nonempty. We are going to prove that for any object A of $\mathcal{A}$, $Sub_{\mathcal{A}}(A)$ and $\theta(\mathcal{P}/A)$ are isomorphic ordered classes. In fact, given a monomorphism $m \colon X \to A$, we consider a $\mathcal{P}$-cover $q \colon Q \to X$, and we get an element in $\theta(\mathcal{P}/A)$ from the composition $m \cdot q$. Conversely, given an object $f \colon Q \to A$ in $\mathcal{P}/A$, the monomorphic part of its regular factorization gives an element in $Sub_{\mathcal{A}}(A)$.

3. $\mathcal{A}$ has coequalizers. Consider a parallel pair (a, b) in $\mathcal{A}$ and its regular factorization:

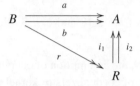

Consider now the equivalence relation $a_1, a_2 \colon A' \rightrightarrows A$ generated by (i_1, i_2), that is, the intersection of all the equivalence relations on A containing (i_1, i_2). Such an intersection exists: by part 2, $Sub_{\mathcal{A}}(A)$ is isomorphic to $\theta(\mathcal{P}/A)$, which is cocomplete because $\mathcal{P}$ has coproducts. Since, by assumption, $\mathcal{A}$ is well powered, $Sub_{\mathcal{A}}(A)$ is a set, and a cocomplete ordered set is also complete. Since $\mathcal{A}$ is exact, (a_1, a_2) has a coequalizer, which is also a coequalizer of (i_1, i_2) and then of (a, b).

4. $\mathcal{A}$ has coproducts. Consider a family of objects $(A_i)_I$ in $\mathcal{A}$. Each object A can be seen as a coequalizer of a pseudoequivalence in $\mathcal{P}$, as in the following diagram, where the first and second columns are coproducts in $\mathcal{P}$ (and then in $\mathcal{A}$; see part 1 above), x_0 and x_1 are the extensions to the coproducts, the bottom row is a coequalizer (which exists by part 3), and σ_i is the extension to the coequalizer:

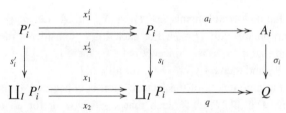

Since coproducts commute with coequalizers, the third column is a coproduct of the family $(A_i)_I$. □

18.6 Corollary *A category is algebraic iff it is exact and has a strong generator $\mathcal{G}$ consisting of finitely presentable regular projectives such that coproducts of objects of $\mathcal{G}$ exist.*

Proof Let $\mathcal{A}$ be an exact category and $\mathcal{G}$ a strong generator consisting of regular projectives. Since a coproduct of regular projectives is regular projective, the full subcategory $\mathcal{P}$ consisting of coproducts of objects from $\mathcal{G}$ is a regular projective cover of $\mathcal{A}$. Following 18.5, it remains to prove that $\mathcal{A}$ is well-powered. Consider an object A and the map

$$\mathcal{F} \colon \Omega(\mathcal{G} \downarrow A) \to Sub_{\mathcal{A}}(A)$$

assigning to a subset $\mathcal{M}$ of $\mathcal{G} \downarrow A$ the subobject of A represented by the monomorphism $s_{\mathcal{M}}: S_{\mathcal{M}} \to A$, where

$$e_{\mathcal{M}}: \coprod_{(G,g) \in \mathcal{M}} G \to A$$

is the canonical morphism whose (G, g)-component is g, and

$$\coprod_{(G,g) \in \mathcal{M}} G \xrightarrow{q_{\mathcal{M}}} S_{\mathcal{M}} \xrightarrow{s_{\mathcal{M}}} A$$

is the regular factorization of $e_{\mathcal{M}}$. We are to prove that $\mathcal{F}$ is surjective so that $\mathrm{Sub}_{\mathcal{A}}(A)$ is a set. For this, consider a monomorphism $m: S \to A$, and let $\mathcal{M}(s)$ be the set of those $(G, g) \in \mathcal{G} \downarrow A$ such that g factorizes through s:

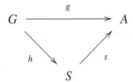

We get the commutative diagram

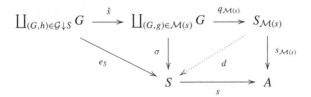

where the (G, h)-components of e_S and $\hat{s}$ are h and g, respectively, and the (G, g)-component of σ is h. By diagonal fill-in, there exists $d: S_{\mathcal{M}(s)} \to S$ such that $d \cdot q_{\mathcal{M}(s)} = \sigma$ and $s \cdot d = s_{\mathcal{M}(s)}$. Such a d is an isomorphism: it is a monomorphism because $s_{\mathcal{M}(s)}$ is, and it is an extremal epimorphism because e_S is. Thus $\mathcal{F}(\mathcal{M}(s)) = s$ and the proof is complete. □

From Propositions 3.18 and 6.22, we know that an algebraic category is exact and locally finitely presentable. The converse is not true because of the lack of projectivity of the generator. In the remaining part of this chapter, we want to state in a precise way the relationship between algebraic categories and exact, locally finitely presentable categories.

18.7 Definition Given a category $\mathcal{B}$, by a *localization* of $\mathcal{B}$ is meant a full reflective subcategory $\mathcal{A}$ whose reflector preserves finite limits. $\mathcal{A}$ is called a *finitary localization* if, moreover, it is closed in $\mathcal{B}$ under filtered colimits.

18.8 Remark More loosely, we speak about localizations of $\mathcal{B}$ as categories equivalent to full subcategories having the preceding property. We use the notation

$$\mathcal{A} \underset{I}{\overset{R}{\leftrightarrows}} \mathcal{B};$$

that is, R is left adjoint to I and I is full and faithful. Let us start with a general lemma.

18.9 Lemma *Consider a full reflection*

$$\mathcal{A} \underset{I}{\overset{R}{\leftrightarrows}} \mathcal{B}.$$

1. *If I preserves filtered colimits and an object $P \in \mathcal{B}$ is finitely presentable, then $R(P)$ is finitely presentable.*
2. *If the reflection is a localization and $\mathcal{B}$ is exact, then $\mathcal{A}$ is exact.*

Proof
1. Use the same argument as in the proof of 6.16.1.

2. Let $r_1, r_2 \colon A' \rightrightarrows A$ be an equivalence relation in $\mathcal{A}$. Its image in $\mathcal{B}$ is an equivalence relation so that it has a coequalizer Q and it is the kernel pair of its coequalizer (because $\mathcal{B}$ is exact):

$$IA' \underset{Ir_2}{\overset{Ir_1}{\rightrightarrows}} IA \overset{q}{\twoheadrightarrow} Q.$$

If we apply the functor R to this diagram, we obtain a coequalizer (because R is a left adjoint) and a kernel pair (because R preserves finite limits),

$$RIA' \simeq A' \underset{r_2}{\overset{r_1}{\rightrightarrows}} A \simeq RIA \overset{Rq}{\twoheadrightarrow} RQ,$$

and this means that (r_1, r_2) is effective. It remains to prove that regular epimorphisms are stable under pullbacks. For this, consider a pullback

$$\begin{array}{ccc} P & \overset{f'}{\longrightarrow} & C \\ {\scriptstyle g'}\downarrow & & \downarrow{\scriptstyle g} \\ A & \underset{f}{\longrightarrow} & B \end{array}$$

in $\mathcal{A}$, with f a regular epimorphism. Consider also its image in $\mathcal{B}$, computed as a two-step pullback of Ig along the regular factorization $m \cdot e$ of If,

$$\begin{array}{ccccc} IP & \xrightarrow{e'} & Q & \xrightarrow{m'} & IC \\ {\scriptstyle Ig'}\downarrow & & \downarrow{\scriptstyle h} & & \downarrow{\scriptstyle Ig} \\ IA & \xrightarrow{e} & E & \xrightarrow{m} & IB \end{array}$$

so that e' is a regular epimorphism. If we apply the functor R to the second diagram, we come back to the original pullback, computed now as a two-step pullback (because R preserves finite limits):

$$\begin{array}{ccccc} P \simeq RIP & \xrightarrow{Re'} & RQ & \xrightarrow{Rm'} & RIC \simeq C \\ {\scriptstyle g'}\downarrow & & \downarrow{\scriptstyle Rh} & & \downarrow{\scriptstyle g} \\ A \simeq RIA & \xrightarrow{Re} & RE & \xrightarrow{Rm} & RIB \simeq B \end{array}$$

Now observe that Rm is a monomorphism (because R preserves finite limits) and also a regular epimorphism (because f is a regular epimorphism and $Rm \cdot Re = f$) so that it is an isomorphism. It follows that Rm' is an isomorphism. Moreover, Re' is a regular epimorphism (because R, being a left adjoint, preserves regular epimorphisms). Finally, f' is a regular epimorphism because $f' = Rm' \cdot Re'$. □

18.10 Theorem *Finitary localizations of algebraic categories are precisely the exact, locally finitely presentable categories.*

Proof Since an algebraic category is exact and locally finitely presentable, necessity follows from 6.16.1 and 18.9. For the sufficiency, let $\mathcal{A}$ be an exact and locally finitely presentable category. Following 6.26, $\mathcal{A}$ is equivalent to $Lex\, \mathcal{T}$, where $\mathcal{T} \simeq \mathcal{A}_{fp}^{op}$, and $Lex\, \mathcal{T}$ is a full reflective subcategory of $Alg\, \mathcal{T}$ closed under filtered colimits (see 6.29). Consider the full subcategory $\mathcal{P}$ of $Alg\, \mathcal{T}$, consisting of regular projective objects. Such an object P is a retract of a coproduct of representable algebras (5.14.2). Since every coproduct is a filtered colimit of its finite subcoproducts, and a finite coproduct of representable algebras is representable (1.13), P is a retract of a filtered colimit of representable algebras. Following 4.3, we have that $\mathcal{P}$ is contained in $Lex\, \mathcal{T}$. Moreover, $\mathcal{P}$ is a regular projective cover of $Alg\, \mathcal{T}$ (5.15). Since, by 16.27, the full inclusion of $\mathcal{P}$ into $Alg\, \mathcal{T}$ is a free exact completion of $\mathcal{P}$ and, by assumption, $Lex\, \mathcal{T}$ is exact, it remains just to prove that the inclusion $\mathcal{P} \to Lex\, \mathcal{T}$ is left covering.

Once this is done, we can apply 16.26 to the following situation:

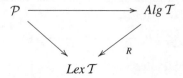

where R is the reflector, and we conclude that R is an exact functor. But the inclusion $\mathcal{P} \to Alg\,\mathcal{T} \simeq \mathcal{P}_{ex}$ is left covering, and $Lex\,\mathcal{T}$ is closed in $Alg\,\mathcal{T}$ under limits so that also the inclusion $\mathcal{P} \to Lex\,\mathcal{T}$ is left covering. □

18.11 Remark To finish this chapter, we observe that Corollary 18.2 can be used to prove the characterization of varieties established in 10.24 without using Birkhoff's variety theorem (10.22). We sketch the argument. Let $\mathcal{T}$ be an algebraic theory and

$$I\colon \mathcal{A} \to Alg\,\mathcal{T}, \quad R \dashv I$$

a regular epireflective subcategory closed under regular quotients and directed unions. From the closedness under regular quotients, it immediately follows that I preserves regular epimorphisms and that in $\mathcal{A}$, equivalence relations are effective, so that $\mathcal{A}$ is exact. To prove that I preserves filtered colimits, consider a functor $F\colon \mathcal{D} \to \mathcal{A}$ with $\mathcal{D}$ filtered, and let $\langle \sigma_d\colon Fd \to B\rangle_{d\in\mathcal{D}}$ be its colimit cocone in $Alg\,\mathcal{T}$. Let

$$Fd \xrightarrow{e_d} Gd \xrightarrow{m_d} B$$

be the regular factorization of σ_d. For any morphism $f\colon d \to d'$ in $\mathcal{D}$, there exists a unique $Gf\colon Gd \to Gd'$ such that $m_{d'} \cdot Gf = m_d$ (use a diagonal fill-in; cf. 0.16). This defines a new functor $G\colon \mathcal{D} \to \mathcal{A}$ (indeed $Gd \in \mathcal{A}$ because it is a regular quotient of $Fd \in \mathcal{A}$), and B is the directed union of the Gds so that $B \in \mathcal{A}$. Following 18.2, I preserves sifted colimits, and then by 6.18, $\mathcal{A}$ is algebraic. Moreover, an algebraic theory $\mathcal{T}_\mathcal{A}$ of $\mathcal{A}$ can be described as follows (cf. 6.16):

$$\mathcal{T}_\mathcal{A}^{op} = \{R(\mathcal{T}(X,-)) \mid X \in \mathcal{T}\}.$$

The functor

$$\mathcal{T} \to \mathcal{T}_\mathcal{A}, \quad X \mapsto R(\mathcal{T}(X,-))$$

preserves finite products (by 1.13) and is surjective on objects. It remains to prove that it is also full: let η be the unit of the adjunction $R \dashv I$, and consider $f\colon R(\mathcal{T}(X,-)) \to R(\mathcal{T}(Z,-))$. Since $\eta_{\mathcal{T}(Z,-)}$ is a regular epimorphism and

$T(X, -)$ is regular projective, there exists $g: T(X, -) \to T(Z, -)$ such that $f \cdot \eta_{T(X,-)} = \eta_{T(Z,-)} \cdot g$. Since η is natural and $\eta_{T(X,-)}$ is an epimorphism, it follows that $Rg = f$. By 10.13, T_A is a quotient of T. Finally,

$$A \simeq Alg\, T_A \simeq Alg\,(T/\sim)$$

so that A is a variety of T-algebras.

Historical remarks

The first systematic study of localizations is from Gabriel (1962), who also, together with Popesco, characterized Grothendieck categories (see Popesco and Gabriel, 1964). This is an ancestor of Theorem 18.10.

Arbitrary (i.e., nonnecessarily finitary) localizations of one-sorted algebraic categories and, more generally, of monadic categories over *Set* are characterized in Vitale (1996) and Vitale (1998). Essential localizations are studied in Adámek et al. (2001b), which generalizes the original result for module categories due to Roos (1965).

One of the results of Lawvere's (1963) thesis is a characterization of one-sorted algebraic categories (cf. Corollary 18.6). The only difference is that in Lawvere's original result, the generator is required to be abstractly finite, a notion that without the other conditions of the characterization theorem is weaker than finitely presentable.

Postscript

In this postscript, we intend to explain somewhat the position our book has in the literature on algebra and category theory, and we want to mention some of the important topics that we decided not to deal with in our book.

One-sorted algebraic theories provide a very convenient formalization, based on the concept of finite product, of the classical concept of "the collection of all algebraic operations" present in a given kind of algebras, for example, in groups or boolean algebras. These theories lead to concrete categories $\mathcal{A}$ of algebras, that is, to categories equipped with a faithful functor $U: \mathcal{A} \to Set$. They can also be used to find an algebraic information present in a given concrete category $\mathcal{A}$: we can form the algebraic theory whose n-ary operations are precisely the natural transformations $U^n \to U$. In the case of groups (and in any one-sorted algebraic category), these "implicit" operations are explicit; that is, they correspond to operations of the theory of groups. But on finite algebras (e.g., finite semigroups), there exist implicit operations that are not explicit, and they are important in the theory of automata (see Almeida, 1994). The passages from one-sorted algebraic theories to one-sorted algebraic categories and back form a duality that is a biequivalence in general. And, as we will see in Appendix C, this passage is an equivalence if we restrict one-sorted algebraic categories to uniquely transportable ones. The latter limitation is caused by a small disadvantage of the formalization based on finite products: a finite product of sets is only isomorphic to the Cartesian product, and thus a finite product preserving functor from a one-sorted algebraic theory to the category of sets is, in general, only isomorphic to a "real" algebra.

General algebraic theories formalize many-sorted algebras. Moreover, the sorting is also variable: whereas classical theories, such as groups and boolean algebras, are one-sorted, we have seen important many-sorted theories for the same algebraic categories. For example the so-called canonical theory of boolean algebras has infinitely many sorts. This canonical theory, which is a foundation of the duality between algebraic categories and algebraic theories, is obtained from any algebraic theory by splitting of idempotents. Its objects do not correspond to finitely generated free algebras but to finitely presentable regular projective algebras. This does not play any role for groups because regular projective groups are free, but it is crucial for boolean algebras and R-modules. Thus the approach of the general algebraic theories goes beyond the traditional algebraic boundaries, for example, it touches homological algebra, in which (regular) projective resolutions are more important than the free ones. The case of chain complexes of R-modules is even more illuminating because finitely presentable regular projective algebras coincide with perfect chain complexes. Forgetting sorts means that we have to consider algebraic categories just as abstract categories and not equipped with a faithful functor to the category of (many)-sorted sets. But there is a way of finding algebraic information present in a given abstract category that takes the dual of the full subcategory of finitely presentable regular projectives. This results in the duality between canonical algebraic theories and algebraic categories.

Algebraic theories immediately lead to sifted colimits, that is, colimits that commute with finite products in sets. In the practice of general algebra, these colimits are mostly reduced to filtered colimits and quotients modulo congruences, and it has taken quite a long time to understand the importance of reflexive coequalizers. Algebraic categories are the free completions of small categories under sifted colimits, which puts them between locally finitely presentable categories and presheaf categories. The fact that algebras are set-valued functors links general algebra with fields like algebraic geometry, where set-valued functors play an important role as sheaves (especially under the influence of Grothendieck; e.g., Artin et al., 1972). Grothendieck toposes, which are the categories of sheaves, can be characterized as the localizations of presheaf categories. In the additive case, one precisely gets Grothendieck categories that are the localizations of categories of modules. Analogously, we have presented a characterization of finitary localizations of algebraic categories. Both the characterization of algebraic categories and their localizations are a combination of exactness properties and a smallness condition (the existence of a suitable generator). Categories satisfying all exactness properties of algebraic categories but no smallness condition form the "equational hull" of varieties where "operations and equations" are not set-like but category-like. For instance, taking a reflexive coequalizer is an operation whose arity is a category, in fact, the reflexive pair. In contrast, the classical construction of forming quotients modulo all congruences cannot be considered as such an operation (which illustrates the importance of reflexive coequalizers in general algebra). The equational hull of varieties was found in Adámek et al. (2001a), which solved the open problem from Lawvere (1969).

One-sorted algebraic theories are closely linked to monads over sets; in fact, they precisely correspond to finitary monads. In the same way, S-sorted algebraic theories correspond to finitary monads over S-sorted sets. See Appendix A. Power (1999) extended this correspondence to certain symmetric monoidal closed categories $\mathcal{V}$ by introducing enriched Lawvere theories and showing that they correspond to finitary $\mathcal{V}$-monads on $\mathcal{V}$. His approach covers some nonalgebraic cases as well, for example, torsion-free Abelian groups that are presented by a finitary monad on Abelian groups. This is caused by the fact that on Abelian groups, contrary to sets (and S-sorted sets), there exist finitary monads that do not preserve sifted colimits. Power's approach can be modified to $\mathcal{V}$-monads preserving sifted colimits (see Lack and Rosický, 2010) on symmetric monoidal closed categories somewhat more restricted. But to deal with monads on symmetric monoidal categories in general, one has to move from Lawvere theories to operads, that is, from finite products to the tensor product (see Boardman and Vogt, 1973; Mac Lane, 1963). This leads to "general algebra over $\mathcal{V}$" (see Loday, 2008; Voronov, 2005). Each operad induces a monad on $\mathcal{V}$, and these operadic monads preserve sifted colimits (see Rezk, 1996).

Given a one-sorted algebraic theory, one can consider its algebras in any category $\mathcal{K}$ having finite products. For instance, for the theory of groups, the algebras in the category of topological spaces are the topological groups, or algebras in the category of smooth manifolds are the Lie groups. But our characterization of algebraic categories strongly depends on the exactness properties of sets and is not applicable in general. Surprisingly, this can be transformed to the homotopy setting, where it reflects the exactness properties of the homotopy category $SSet$ of simplicial sets (which is equivalent to the classical homotopy category of topological spaces). One considers homotopy algebras of $\mathcal{T}$: they are functors $A: \mathcal{T} \to SSet$, preserving finite products up to homotopy, which means that the canonical maps

$$A(t_1 \times \cdots \times t_n) \to A(t_1) \times \cdots \times A(t_n)$$

are not isomorphisms but weak equivalences. An analogy of sifted colimits emerges, but reflexive coequalizers are replaced by the homotopy colimits of simplicial objects (a reflexive pair is the 2-truncation of a simplicial object). The resulting characterization of homotopy varieties can be found in Rosický (2007); independently, it was presented by Lurie (2009) using the language of quasicategories of Joyal (2008).

Appendix A
Monads

An important aspect of algebraic categories that has not yet been treated in this book are monads. The aim of this appendix is to give a short introduction to monads on a category $\mathcal{K}$ and then to explain how finitary monads for $\mathcal{K} = Set$ precisely yield one-sorted algebraic theories, and for $\mathcal{K} = Set^S$, the S-sorted ones.

The word *monad* stems from *monoid*: recall that a monoid in a category $\mathcal{K}$ is an object M together with a morphism $m \colon M \times M \to M$, which (1) is associative, that is, the square

$$\begin{array}{ccc} M \times M \times M & \xrightarrow{m \times \mathrm{id}_M} & M \times M \\ {\scriptstyle \mathrm{id}_M \times m} \downarrow & & \downarrow {\scriptstyle m} \\ M \times M & \xrightarrow{m} & M \end{array}$$

commutes, and (2) has a unit, that is, a morphism $e \colon 1 \to M$ such that the triangles

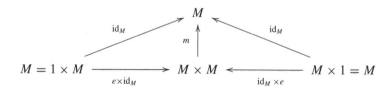

commute.

207

A.1 Definition

1. A *monad* $\mathbb{M}$ on a category $\mathcal{K}$ consists of an endofunctor M on $\mathcal{K}$ and natural transformations (1) $\mu\colon MM \to M$ (monad multiplication) and (2) $\eta\colon \mathrm{Id}_{\mathcal{K}} \to M$ (monad unit) such that the diagrams

$$\begin{array}{ccc} MMM & \xrightarrow{M\mu} & MM \\ {\scriptstyle \mu M}\downarrow & & \downarrow{\scriptstyle \mu} \\ MM & \xrightarrow{\mu} & M \end{array} \qquad (A.1)$$

and

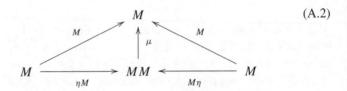
(A.2)

commute.
2. The monad is called *finitary* if M is a finitary functor.

A.2 Example

1. The functor $M\colon Set \to Set$ given by

$$MX = X + 1$$

carries the obvious structure of a monad: the unit is the coproduct injection $\eta_X\colon X \to X + 1$, and the multiplication $\mu_X\colon X + 1 + 1 \to X + 1$ merges the two copies of 1 to a single copy.

2. The *word monad* on *Set* assigns to every set X the set

$$MX = X^*$$

of all words on it, that is, the (underlying set of the) free monoid on X. This yields an endofunctor on *Set* together with natural transformations $\eta_X\colon X \to X^*$, the formation of one-letter words, and $\mu_X\colon (X^*)^* \to X^*$ given by concatenation of words.

In the notation of 12.6, we have two natural transformations

$$\eta\colon \mathrm{Id} \to U_H F_H = H^*$$
$$\varphi\colon HH^* \to H^*,$$

yielding a natural transformation
$$\psi = \varphi \cdot H\eta \colon H \to H^*.$$

We will see in Corollary A.27 that it has the universal property explaining the name free monad.

A.3 Example The basic example of a monad is that induced by any adjoint situation

$$\mathcal{A} \underset{U}{\overset{F}{\rightleftarrows}} \mathcal{K}$$

where $F \dashv U$. Let $\eta \colon \mathrm{Id}_\mathcal{K} \to UF$ and $\varepsilon \colon FU \to \mathrm{Id}_\mathcal{A}$ denote the unit and counit of the adjunction. Recall the equalities

$$U = U\varepsilon \cdot \eta U \quad \text{and} \quad F = \varepsilon F \cdot F\eta \tag{A.3}$$

characterizing adjoint situations. Then, for the endofunctor

$$M = UF \colon \mathcal{K} \to \mathcal{K},$$

we have the natural transformation

$$\mu = U\varepsilon F \colon MM = UFUF \to UF = M, \tag{A.4}$$

which, together with the unit $\eta \colon \mathrm{Id}_\mathcal{K} \to M$, forms a monad on $\mathcal{K}$. In fact, the commutativity of the two triangles (A.2) follows from Equations (A.3), and the commutativity of square (A.1) follows from the naturality of ε:

$$\varepsilon \cdot FU\varepsilon = \varepsilon \cdot \varepsilon FU, \tag{A.5}$$

yielding

$$\mu \cdot M\mu = U(\varepsilon \cdot FU\varepsilon)F = U(\varepsilon \cdot \varepsilon FU)F = \mu \cdot \mu M.$$

Observe that whenever U is a finitary functor, this monad is finitary because F, being a left adjoint, always preserves filtered colimits.

A.4 Example

1. Every one-sorted algebraic category $U \colon \mathcal{A} \to \mathit{Set}$ defines a monad on Set assigning to every set X the free algebra generated by it. In other words, this is the monad induced by the adjunction $F \dashv U$, as in A.3, where F is the free-algebra functor of 11.21. Since U preserves filtered colimits by 11.8, all these monads are finitary.
2. Analogously, every S-sorted algebraic category defines a finitary monad on Set^S.

3. Recall that for every finitary endofunctor H of *Set*, free H-algebras exist, giving a left adjoint $F_H \colon Set \to H\text{-}Alg$ (see 12.7). The corresponding monad H^* on *Set* is called the *free monad on H*: it assigns to every set X the free H-algebra on X. For example, if H is the polynomial functor of a signature Σ, the free monad is the monad of Σ-terms, see 13.1.

A.5 Remark Recall from 12.1 the category $M\text{-}Alg$ of M-algebras for the endofunctor M of $\mathcal{K}$. If $\mathbb{M}$ is the monad induced by an adjoint situation $F \dashv U$, as in A.3, then every object A of $\mathcal{A}$ yields a canonical M-algebra on $X = UA$: put

$$x = U\varepsilon_A \colon MX = UFUA \to UA = X.$$

This algebra has the property that the triangle

$$\begin{array}{ccc} & X & \\ {\scriptstyle \eta_X} \downarrow & \searrow {\scriptstyle \mathrm{id}_X} & \\ MX & \xrightarrow{\ x\ } & X \end{array} \qquad (A.6)$$

commutes; see Equations (A.3). Also, the square

$$\begin{array}{ccc} MMX & \xrightarrow{\ \mu_X\ } & MX \\ {\scriptstyle Mx} \downarrow & & \downarrow {\scriptstyle x} \\ MX & \xrightarrow{\ x\ } & X \end{array} \qquad (A.7)$$

commutes; see Equation (A.5). This leads to the following

A.6 Definition An *Eilenberg–Moore algebra* for a monad $\mathbb{M} = (M, \mu, \eta)$ on $\mathcal{K}$ is an algebra (X, x) for M such that the diagrams (A.6) and (A.7) commute. The full subcategory of $M\text{-}Alg$ formed by all Eilenberg–Moore algebras is denoted by

$$\mathcal{K}^{\mathbb{M}}.$$

A.7 Remark The Eilenberg–Moore category $\mathcal{K}^{\mathbb{M}}$ is considered a concrete category on $\mathcal{K}$ via the faithful functor

$$U_{\mathbb{M}} \colon \mathcal{K}^{\mathbb{M}} \to \mathcal{K}, \quad (X, x) \mapsto X.$$

It is easy to verify that this concrete category is uniquely transportable (same argument as in 13.17.3).

A.8 Example

1. For every category $\mathcal{K}$, we have the trivial monad $\mathbb{I}d = (\mathrm{Id}_\mathcal{K}, \mathrm{id}, \mathrm{id})$. The only Eilenberg–Moore algebras are $\mathrm{id}_X\colon X \to X$. Thus $\mathcal{K}^\mathbb{I} \simeq \mathcal{K}$.
2. For the monad $MX = X + 1$ of A.2.1, an Eilenberg–Moore algebra is a pointed set: given $x\colon X + 1 \to X$ for which Triangle (A.6) commutes, the left-hand component of x is id_X, and thus x just chooses an element $1 \to X$. Here Square (A.7) always commutes. Homomorphisms are functions preserving the choice of element. In short, $Set^\mathbb{M}$ is the category of pointed sets.
3. For the word monad A.2.2, the category $Set^\mathbb{M}$ is essentially the category of monoids. In fact, given an Eilenberg–Moore algebra $x\colon X^* \to X$, Triangle (A.6) states that the response to one-letter words is trivial: $x(a) = a$, and Square (A.7) states that for words of length larger than 2, the response is given by the binary operation

$$a_1 * a_2 = x(a_1 a_2).$$

In fact, for example, with length 3, we get

$$x(a_1 a_2 a_3) = x(a_1(a_2 a_3)) = a_1 * (a_2 * a_3)$$

as well as

$$x(a_1 a_2 a_3) = x((a_1 a_2) a_3) = (a_1 * a_2) * a_3.$$

Thus $*$ is an associative operation. Square (A.7) also states that the response of x to the empty word is a unit for x.

Conversely; every monoid defines an Eilenberg–Moore algebra, see Remark A.5. The monoid homomorphisms are easily seen to be precisely the homomorphisms in $Set^\mathbb{M}$. Thus $Set^\mathbb{M}$ is isomorphic to the category of monoids.

A.9 Example: Free Eilenberg–Moore algebras For every monad $\mathbb{M}$, the M-algebra

$$(MX, \mu_X\colon MMX \to MX)$$

is an Eilenberg–Moore algebra: the commutativity of the diagrams (A.6) and (A.7) follow from the definition of monad. This algebra is free with respect to $\eta_X\colon X \to MX$. In fact, given an Eilenberg–Moore algebra (Z, z) and a morphism $f\colon X \to Z$ in $\mathcal{K}$, the unique homomorphism extending f is $\overline{f} = z \cdot Mf$:

1. $\overline{f} = z \cdot Mf$ is a homomorphism: the diagram

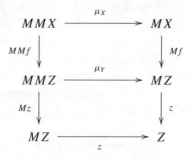

commutes due to Equation (A.5) and the naturality of μ.
2. Conversely, if $\overline{f}\colon (MX, \mu_X) \to (Z, z)$ is a homomorphism, then $f = \overline{f} \cdot \eta_X$ implies that $\overline{f} = z \cdot Mf$:

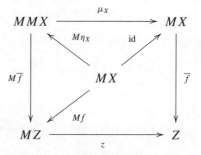

A.10 Corollary *Every monad is induced by some adjoint situation.*

In fact, given a monad $\mathbb{M}$ on a category $\mathcal{K}$, we have the preceding adjoint situation
$$\mathcal{K}^{\mathbb{M}} \underset{U_{\mathbb{M}}}{\overset{F_{\mathbb{M}}}{\leftrightarrows}} \mathcal{K},$$
where $F_{\mathbb{M}}$ is the free-algebra functor
$$F_{\mathbb{M}} X = (MX, \mu_X).$$
It is defined on morphisms by $F_{\mathbb{M}} f = Mf$. Thus the monad induced by the adjunction $F_{\mathbb{M}} \dashv U_{\mathbb{M}}$ has the underlying endofunctor $U_{\mathbb{M}} \cdot F_{\mathbb{M}} = M$ and the unit η (recall the universal morphisms η_X from A.9). We need to verify that for the counit ε of the adjoint situation, we have
$$\mu = U_{\mathbb{M}} \,\varepsilon\, F_{\mathbb{M}}.$$

In fact, the component of ε at an Eilenberg–Moore algebra (X, x) is the unique homomorphism

$$\varepsilon_{(X,x)}: (MX, \mu_X) \to (X, x)$$

with $U_\mathbb{M}\varepsilon_{(X,x)} \cdot \eta_X = \mathrm{id}_X$. But because of the diagrams (A.6) and (A.7), the morphism x carries such a homomorphism. Therefore $U_\mathbb{M}\varepsilon_{(X,x)} = x$ for all algebras (X, x). In particular,

$$(U_\mathbb{M} \varepsilon\, F_\mathbb{M})_X = U_\mathbb{M}\varepsilon_{(MX,\mu_X)} = \mu_X.$$

A.11 Definition For every adjoint situation

$$\mathcal{A} \underset{U}{\overset{F}{\rightleftarrows}} \mathcal{K} \quad \text{with } F \dashv U,$$

let $\mathbb{M}$ be the monad of A.3. The *comparison functor* is the functor

$$K: \mathcal{A} \to \mathcal{K}^\mathbb{M}$$

that assigns to every object A the Eilenberg–Moore algebra

$$KA = (UA, U\varepsilon_A)$$

of A.5. The definition of K on morphisms $f: A \to B$ uses the naturality of ε, which shows that $Kf = Uf$ is a homomorphism:

$$\begin{array}{ccc} MUA & \xrightarrow{U\varepsilon_A} & UA \\ {\scriptstyle MUf} \downarrow & & \downarrow {\scriptstyle Uf} \\ MUB & \xrightarrow[U\varepsilon_B]{} & UB \end{array}$$

A.12 Remark The comparison functor $K: \mathcal{A} \to \mathcal{K}^\mathbb{M}$ of A.11 is the unique functor such that $U_\mathbb{M} \cdot K = U$ and $K \cdot F = F_\mathbb{M}$.

A.13 Example

1. For the concrete category of monoids $U: \mathit{Mon} \to \mathit{Set}$, the comparison functor $K: \mathit{Mon} \to \mathit{Set}^\mathbb{M}$ is an isomorphism. The inverse K^{-1} was described in A.8.2 on objects and acts trivially on morphisms: $K^{-1}f = f$.
2. The concrete category $U: \mathit{Pos} \to \mathit{Set}$ of partially ordered sets yields the trivial monad $\mathbb{Id} = (\mathrm{Id}_{\mathit{Set}}, \mathrm{id}, \mathrm{id})$ of A.8: recall that the left adjoint of U

assigns to every set X the discrete order on the same set. For this monad, we have an isomorphism between *Set* and $Set^{\mathbb{M}}$, and the comparison functor is then simply the forgetful functor U.

3. For the free monad H^* on H (see A.4.3), the Eilenberg–Moore category is concretely isomorphic to H-*Alg*. Indeed, the functor $J: Set^{H^*} \to H\text{-}Alg$ taking an Eilenberg–Moore algebra $x: H^*X \to X$ to the H-algebra obtained by composing with ψ_X (see A.4.3) is easily seen to be a concrete isomorphism.

A.14 Definition A concrete category $(\mathcal{A}, U)$ on $\mathcal{K}$ is *monadic* if U has a left adjoint F and the comparison functor $K: \mathcal{A} \to \mathcal{K}^{\mathbb{M}}$ is an isomorphism for the monad $\mathbb{M}$ induced by $F \dashv U$.

A.15 Remark

1. The fact that $(\mathcal{A}, U)$ is monadic does not depend on the choice of the left adjoint of U. Indeed, if F' is another left adjoint, then the canonical natural isomorphism $F \simeq F'$ induces an isomorphism of monads $\mathbb{M} \simeq \mathbb{M}'$ (see A.24 for the notion of monad morphism), where $\mathbb{M}'$ is the monad induced by the adjunction $U \dashv F'$. As we will see in A.25, this implies that $\mathcal{K}^{\mathbb{M}}$ and $\mathcal{K}^{\mathbb{M}'}$ are concretely isomorphic.
2. In other words, monadic concrete categories are precisely those that, up to concrete isomorphism, have the form $\mathcal{K}^{\mathbb{M}}$. It is not surprising, then, that monoids are an example of a monadic concrete category and posets are not.

A.16 Definition A coequalizer in a category $\mathcal{K}$ is called *absolute* if every functor with domain $\mathcal{K}$ preserves it.

A.17 Example For every Eilenberg–Moore algebra (X, x), we have an absolute coequalizer

$$MMX \underset{\mu_X}{\overset{Mx}{\rightrightarrows}} MX \xrightarrow{x} X$$

in $\mathcal{K}$. In fact, x merges the parallel pair by Square (A.7), and moreover, the morphisms η_X and η_{MX} are easily seen to fulfill the following equations: $\mu_X \cdot \eta_{MX} = \text{id}_{MX}$, $\eta_X \cdot x = \text{id}_X$, and $Mx \cdot \eta_{MX} = \eta_X \cdot x$. It is easy to derive from these equations that x is a coequalizer of Mx and μ_X. Since every functor $G: \mathcal{K} \to \mathcal{L}$ preserves the preceding equations, it follows that Gx is a coequalizer of GMx and $G\mu_X$.

A.18 Beck's theorem: *Characterization of monadic categories* A concrete category $(\mathcal{A}, U)$, with U a right adjoint, is monadic iff (1) it is uniquely transportable and (2) $\mathcal{A}$ has coequalizers of all reflexive pairs f, g such that Uf, Ug have an absolute coequalizer; and U preserves these coequalizers.

A proof of A.18 can be found in Mac Lane (1998, chap. 6, sec. 7). The reader has just to observe that the parallel pairs of morphisms used in that proof are reflexive, and U creates the coequalizers involved in condition (2) because it is amnestic and conservative.

A.19 Proposition *Every equational category (13.10.2) is monadic.*

Proof Condition 1 of A.18 follows from 13.17 and condition 2 from 13.11 and 11.8: the forgetful functor is algebraic and thus it preserves reflexive coequalizers. □

A.20 Example

1. Pointed sets, monoids, groups, abelian groups, and so on with their forgetful functors, are monadic.
2. For a one-sorted algebraic theory $(\mathcal{T}, T)$, the concrete category $(Alg\,\mathcal{T}, Alg\,T)$ in general is not monadic, as Example 11.7 shows: $Alg\,T_{ab}$ is not amnestic, whereas $U_\mathbb{M}$ always is. What remains true is that $(Alg\,\mathcal{T}, Alg\,T)$ is *pseudomonadic*, as we will see in Proposition C.4 (see Appendix C).

A.21 Theorem *Equational categories are up to concrete isomorphism precisely the categories $Set^\mathbb{M}$ of Eilenberg–Moore algebras for finitary monads $\mathbb{M}$ on Set.*

Proof In fact, every equational category is, by A.19, concretely isomorphic to $Set^\mathbb{M}$, where $\mathbb{M}$ is the monad of its free algebras. Conversely, given a finitary monad $\mathbb{M} = (M, \mu, \eta)$ on Set, we know from 13.23 that M-Alg is concretely isomorphic to an equational category. Therefore it is sufficient to prove that $Set^\mathbb{M}$ is closed in M-Alg under products, subobjects, and regular quotients. Then the result follows from Birkhoff's variety theorem in the form 13.22.

1. For products; let $(X, x) = \prod_{i \in I}(X_i, x_i)$, where each (X_i, x_i) is an Eilenberg–Moore algebra. For the algebra (X, x) Triangle (A.6) commutes because the projections $(\pi_i)_{i \in I}$ are a limit cone, thus collectively monomorphic,

and the diagram

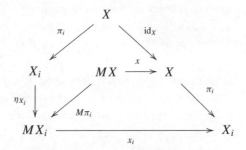

commutes for every i. Thus $x_i \cdot \eta_{X_i} = \mathrm{id}$ implies $x \cdot \eta_X = \mathrm{id}$. For the algebra (X, x) Square (A.7) commutes for similar reasons:

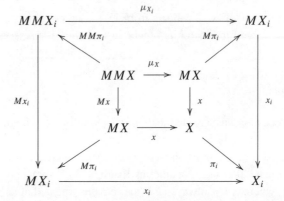

2. For subalgebras $m: (X, x) \to (Z, z)$ of Eilenberg–Moore algebras (Z, z), in the diagram

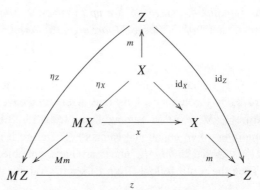

the outward triangle commutes, and all parts, except the middle triangle, also commute. Thus so does the middle triangle since m is a monomorphism. Therefore Triangle (A.6) commutes for (X, x). The proof of Square (A.7) is analogous.

3. For regular quotients $e \colon (Z, z) \to (X, x)$ of Eilenberg–Moore algebras (Z, z), in the diagram

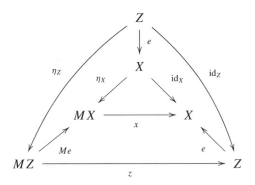

again, all parts, except the middle triangle, commute. Since e is an epimorphism, so does the middle triangle. Therefore Triangle (A.6) commutes. For Square (A.7), use the analogous argument plus the fact that M preserves epimorphisms (because they split in *Set*). □

A.22 Corollary *One-sorted algebraic categories are up to concrete equivalence precisely the categories $Set^{\mathbb{M}}$ of Eilenberg–Moore algebras for finitary monads $\mathbb{M}$ on Set.*

In fact, this follows from A.21 and 13.11.

A.23 Corollary *For every finitary monad $\mathbb{M}$ on Set, the category $Set^{\mathbb{M}}$ is cocomplete, and the forgetful functor $U_{\mathbb{M}} \colon Set^{\mathbb{M}} \to Set$ preserves sifted colimits.*

In fact, the category $Set^{\mathbb{M}}$ is equational by A.21 and then one-sorted algebraic by 13.11. Use now 4.5 and 11.9.

A.24 Definition Let $\mathbb{M} = (M, \mu, \eta)$ and $\mathbb{M}' = (M', \mu', \eta')$ be monads on a category $\mathcal{K}$. A *monad morphism* from $\mathbb{M}$ to $\mathbb{M}'$ is a natural transformation $\rho \colon M \to M'$ such that the diagrams

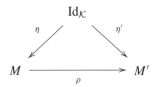

and

$$
\begin{array}{ccc}
MM & \xrightarrow{M\rho} & MM' & \xrightarrow{\rho M'} & M'M' \\
\mu \downarrow & & & & \downarrow \mu' \\
M & & \xrightarrow{\rho} & & M'
\end{array}
$$

commute.

A.25 Proposition *Every monad morphism $\rho: \mathbb{M} \to \mathbb{M}'$ induces a concrete functor*

$$H_\rho: \mathcal{K}^{\mathbb{M}'} \to \mathcal{K}^{\mathbb{M}},$$

assigning to every Eilenberg–Moore algebra (X, x) for $\mathbb{M}'$ the algebra $(X, x \cdot \rho_X)$ for $\mathbb{M}$. Conversely, given a concrete functor

$$
\begin{array}{ccc}
\mathcal{K}^{\mathbb{M}'} & \xrightarrow{H} & \mathcal{K}^{\mathbb{M}} \\
& \searrow^{U_{\mathbb{M}'}} \quad \swarrow_{U_\mathbb{M}} & \\
& \mathcal{K} &
\end{array}
$$

there exists a unique monad morphism $\rho: \mathbb{M} \to \mathbb{M}'$ with $H = H_\rho$.

In fact, for a free Eilenberg–Moore algebra $(M'A, \mu'_A)$, the algebra $H(M'A, \mu'_A)$ has the form $(M'A, \sigma_A: MM'A \to M'A)$. We get a monad morphism

$$\rho: \mathbb{M} \to \mathbb{M}', \quad \rho_A: MA \xrightarrow{M\eta'_A} MM'A \xrightarrow{\sigma_A} M'A.$$

A full proof can be found in Borceux (1994, vol. 2, proposition 4.5.9).

A.26 Corollary *The category of finitary monads on Set and monad morphisms is dually equivalent to the category of finitary monadic categories on Set and concrete functors.*

In fact, this follows from A.25.

A.27 Corollary *The free monad H^* on a finitary endofunctor H (see A.4.3) is indeed free on H: for every finitary monad $\mathbb{M} = (M, \mu, \eta)$ and every natural transformation $\alpha: H \to M$, there exists a unique monad morphism $\alpha^*: H^* \to M$ with $\alpha = \alpha^* \cdot \psi$.*

In fact, recall from A.13.3 that H-Alg is concretely isomorphic to Set^{H^*}, and use, in place of α^*, the concrete functor from $Set^{\mathbb{M}}$ to H-Alg, assigning to every Eilenberg–Moore algebra $x\colon MX \to X$ the algebra $x \cdot \alpha_X\colon HX \to X$.

A.28 Remark We observed in A.10 that every monad $\mathbb{M}$ is induced by an adjoint situation using the Eilenberg–Moore algebras. There is another way to induce $\mathbb{M}$: the construction of the Kleisli category of $\mathbb{M}$. This is, as we note later, just the full subcategory of $\mathcal{K}^{\mathbb{M}}$ on all free algebras.

A.29 Definition The *Kleisli category* of a monad $\mathbb{M}$ is the category

$$\mathcal{K}_{\mathbb{M}}$$

with the same objects as $\mathcal{K}$ and with morphisms from X to Z given by morphisms $f\colon X \to MZ$ in $\mathcal{K}$:

$$\mathcal{K}_{\mathbb{M}}(X, Z) = \mathcal{K}(X, MZ).$$

The identity morphisms are η_X, and the composition of two morphisms $f \in \mathcal{K}_{\mathbb{M}}(X, Z)$ and $g \in \mathcal{K}_{\mathbb{M}}(Z, W)$ is given by the following composition in $\mathcal{K}$:

$$X \xrightarrow{f} MZ \xrightarrow{Mg} MMW \xrightarrow{\mu_W} MW.$$

A.30 Example For the monad $MX = X + 1$ of A.2.1, the Kleisli category is the category of sets and partial functions. A partial function from X to Z is represented as a (total) function from X to $Z + 1$.

A.31 Notation For every monad $\mathbb{M}$, we denote (1) by $K_{\mathbb{M}}\colon \mathcal{K}_{\mathbb{M}} \to \mathcal{K}^{\mathbb{M}}$ the functor that assigns to $X \in \mathcal{K}_{\mathbb{M}}$ the free Eilenberg–Moore algebra (MX, μ_X) and to $f \in \mathcal{K}_{\mathbb{M}}(X, Z)$ the morphism $MX \xrightarrow{Mf} MMZ \xrightarrow{\mu_Z} MZ$, and (2) by $J_{\mathbb{M}}\colon \mathcal{K} \to \mathcal{K}_{\mathbb{M}}$ the functor that is the identity map on objects and that to every morphism $u\colon X \to Z$ of $\mathcal{K}$ assigns

$$X \xrightarrow{u} Z \xrightarrow{\eta_Z} MZ.$$

A.32 Lemma *The functor $J_{\mathbb{M}}$ is a left adjoint of*

$$\mathcal{K}_{\mathbb{M}} \xrightarrow{K_{\mathbb{M}}} \mathcal{K}^{\mathbb{M}} \xrightarrow{U_{\mathbb{M}}} \mathcal{K},$$

and $\mathbb{M}$ is the monad induced by the adjunction $J_\mathbb{M} \dashv U_\mathbb{M} \cdot K_\mathbb{M}$. The functor $K_\mathbb{M}$ is the corresponding comparison functor; it is full and faithful.

Proof $J_\mathbb{M}$ is a left adjoint of $U_\mathbb{M} \cdot K_\mathbb{M}$ with unit

$$\eta_X \colon X \to MX = U_\mathbb{M} K_\mathbb{M} J_\mathbb{M} X$$

and counit given by the morphism in $\mathcal{K}_\mathbb{M}(J_\mathbb{M} U_\mathbb{M} K_\mathbb{M} X, X)$, which is $\varepsilon = \mathrm{id}_{MX}$ in $\mathcal{K}$. The two axioms

$$\varepsilon J_\mathbb{M} \cdot J_\mathbb{M} \eta = J_\mathbb{M} \quad \text{and} \quad U_\mathbb{M} K_\mathbb{M} \varepsilon \cdot \eta U_\mathbb{M} K_\mathbb{M} = U_\mathbb{M} K_\mathbb{M}$$

are easy to check. □

A.33 Theorem *Given monads $\mathbb{M}$ and $\mathbb{M}'$ on a category $\mathcal{K}$, there is a bijective correspondence between monad morphisms $\rho \colon \mathbb{M} \to \mathbb{M}'$ and functors $G \colon \mathcal{K}_\mathbb{M} \to \mathcal{K}_{\mathbb{M}'}$ for which the triangle*

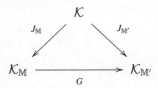

commutes.

We will see in the proof that the bijective correspondence assigns to every monad morphism $\rho \colon \mathbb{M} \to \mathbb{M}'$ the functor $\hat{\rho} \colon \mathcal{K}_\mathbb{M} \to \mathcal{K}_{\mathbb{M}'}$, which is the identity map on objects, and assigns to $p \colon X \to MZ$ in $\mathcal{K}_\mathbb{M}(X, Z)$ the value

$$X \xrightarrow{p} MZ \xrightarrow{\rho_Z} M'Z.$$

Proof

1. The functor $\hat{\rho}$ is well defined: preservation of identity morphisms follows from $\rho_X \cdot \eta_X = \eta'_X$. Preservation of composition follows from the commutativity of the following diagram, where $p \colon X \to MW$ and $q \colon W \to MZ$ are

arbitrary morphisms:

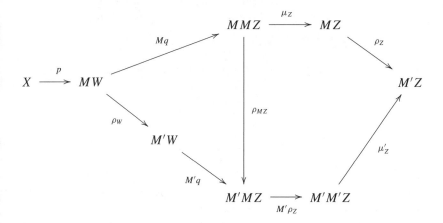

2. The equality $\hat{\rho} \cdot J_{\mathbb{M}} = J_{\mathbb{M}'}$ follows from $\rho \cdot \eta = \eta'$.

3. If ρ, σ are different monad morphisms, then $\hat{\rho} \neq \hat{\sigma}$ due to the fact that the component ρ_X is obtained from $\hat{\rho}$ by $\hat{\rho}(\mathrm{id}_{MX}) = \rho_X$.

4. Let $G \colon \mathcal{K}_{\mathbb{M}} \to \mathcal{K}_{\mathbb{M}'}$ be a functor such that $G \cdot J_{\mathbb{M}} = J_{\mathbb{M}'}$. Observe that this condition tells us that G is the identity map on objects. The identity morphism $\mathrm{id}_{MX} \colon MX \to MX$ can be seen as a morphism $\varepsilon_X \colon MX \to X$ in $\mathcal{K}_{\mathbb{M}}$. Applying G, we get a morphism $G\varepsilon_X \colon MX \to X$ in $\mathcal{K}_{\mathbb{M}'}$, that is, a morphism $MX \to M'X$ in $\mathcal{K}$ that we denote by ρ_X. We claim that ρ_X is the component at X of a monad morphism $\rho \colon \mathbb{M} \to \mathbb{M}'$ with $\hat{\rho} = G$.

The fact that G is a functor means that

(1) $G\eta_X = \eta'_X$

(preservation of identity morphisms); given $p \colon X \to MW$ and $q \colon W \to MZ$ in $\mathcal{K}$, we have

(2) $$G(X \xrightarrow{p} MW \xrightarrow{Mq} MMZ \xrightarrow{\mu_Z} MZ)$$
$$= (X \xrightarrow{Gp} M'W \xrightarrow{M'Gq} M'M'Z \xrightarrow{\mu'_Z} M'Z)$$

(preservation of composition); and for every $u \colon X \to W$ in $\mathcal{K}$, we have

(3) $G(X \xrightarrow{u} W \xrightarrow{\eta_W} MW) = (X \xrightarrow{u} W \xrightarrow{\eta'_W} M'W)$

(due to $G \cdot J_{\mathbb{M}} = J_{\mathbb{M}'}$). This implies for all $u \colon X \to W$ in $\mathcal{K}$ the equation

(4) $GMu = Gu \cdot \rho_X$

since we apply (2) to $p = \mathrm{id}_{MX}$ and $q = \eta_W \cdot u$ (thus the previous objects X, W, and Z are now MX, X, and W, respectively, and $Gq = \eta'_W \cdot u$ by (3)) and use (A.2) for M and M'. Also, for every $q \colon W \to MZ$, we have

(5) $G(q \cdot u) = Gq \cdot u$

by applying (2) and (3) to $p = \eta_W \cdot u$. From this, we derive the naturality of ρ:

(6) $\rho_W \cdot Mv = M'v \cdot \rho_X$ for all $v \colon X \to W$,

since (5) yields for $q = \mathrm{id}_{MW}$ and $u = Mv$

$$\rho_W \cdot Mv = GMv,$$

and then we apply (4). We are ready to prove that ρ is a monad morphism. The equality $\rho \cdot \eta = \eta'$ follows from (5) by $u = \eta_W$ and $q = \mathrm{id}_{MW}$, and the equality $\rho \cdot \mu = \mu' \cdot \rho M' \cdot M\rho$ follows from (2) by $p = \mathrm{id}_{MMX}$ and $q = \mathrm{id}_{MX}$; this proves that the right-hand side is equal to $G\mu_X$, whereas the left-hand side is $G\mu_X$ by (5) applied to $q = \mu_X$ and $u = \mathrm{id}_{MMX}$. Thus

$$\rho \colon \mathbb{M} \to \mathbb{M}'$$

is a monad morphism. Finally, we need to prove $\hat{\rho} = G$; that is, for every $p_0 \colon X \to MW$, we have that

$$Gp_0 = G\varepsilon_X \cdot p_0,$$

and for this, apply (5) to $u = p_0$ and $q = \mathrm{id}_{MW}$. □

A.34 Remark Let $\mathbb{M}$ be a finitary monad on *Set*. Then the functor M is essentially determined by its domain restriction to $\mathcal{N}^{op}$ (the full subcategory of natural numbers) since Set is a free completion Ind $\mathcal{N}^{op}$ of $\mathcal{N}^{op}$ under sifted colimits, see 4.13. Also, the natural transformations η and μ are uniquely determined by their components η_n and μ_n for natural numbers n.

This leads us to the following restriction of the Kleisli category.

A.35 Notation For every finitary monad $\mathbb{M}$ on *Set*, we denote by

$$Set^f_{\mathbb{M}}$$

the full subcategory of the Kleisli category $\mathcal{K}_{\mathbb{M}}$ on all natural numbers and by

$$J^f_{\mathbb{M}} \colon \mathcal{N}^{op} \to Set^f_{\mathbb{M}}$$

the domain-codomain restriction of $J_{\mathbb{M}}$.

A.36 Corollary *Given finitary monads $\mathbb{M}$ and $\mathbb{M}'$ on Set, there is a bijective correspondence between monad morphisms $\rho \colon \mathbb{M} \to \mathbb{M}'$ and those functors*

$G: Set^f_{\mathbb{M}} \to Set^f_{\mathbb{M}'}$, for which the triangle

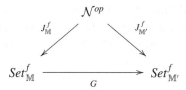

commutes. It assigns to ρ the functor

$$\hat{\rho}^f: (n \xrightarrow{p} Mk) \mapsto (n \xrightarrow{p} Mk \xrightarrow{\rho_k} M'k).$$

The proof is completely analogous to that of A.33. All we need to notice is that a monad morphism $\rho: \mathbb{M} \to \mathbb{M}'$ is uniquely determined by its components ρ_n, $n \in \mathcal{N}^{op}$. This follows once again from 4.13 applied to $Y_{\mathcal{N}}: \mathcal{N}^{op} \to Set$.

In 11.35, we defined the 2-category Th^1 of one-sorted algebraic theories. In the following theorem, we consider it as a category; that is, we forget the 2-cells.

A.37 Theorem *The category Th^1 of one-sorted algebraic theories is equivalent to the category of finitary monads on Set.*

Proof Denote by *FMon* the category of finitary monads and monad morphisms. Every object $\mathbb{M}$ defines a one-sorted theory by dualizing $J^f_{\mathbb{M}}$ of A.35:

$$(J^f_{\mathbb{M}})^{op}: \mathcal{N} \to (Set^f_{\mathbb{M}})^{op}.$$

In fact, since $J_{\mathbb{M}}: Set \to Set_{\mathbb{M}}$ is a left adjoint, it preserves coproducts. In other words, coproducts are the same in *Set* and in $Set_{\mathbb{M}}$. The same, then, holds for finite coproducts in the full subcategories $\mathcal{N}^{op}$ and $Set^f_{\mathbb{M}}$, respectively. Therefore the identity-on-objects functor $(J^f_{\mathbb{M}})^{op}$ preserves finite products, and we obtain a one-sorted theory

$$E(\mathbb{M}) = \left((Set^f_{\mathbb{M}})^{op}, (J^f_{\mathbb{M}})^{op}\right).$$

This yields a functor

$$E: FMon \to Th^1,$$

which, to every monad morphism $\rho: \mathbb{M} \to \mathbb{M}'$ between finitary monads, assigns

$$E(\rho) = (\hat{\rho}^f)^{op}.$$

The commutative triangle of A.36 tells us that this is a morphism of one-sorted theories. It is easy to verify that E is a well-defined functor.

Next, E is full and faithful because of the bijection in A.36. Finally, E is essentially surjective. Recall from 11.21 that for every one-sorted theory $(\mathcal{T}, T)$, we have a left adjoint F_T to the forgetful functor $Alg\, T$ (of evaluation at 1) such that

$$F_T(n) = \mathcal{T}(n, -) \quad \text{and} \quad F_T(\pi_i^n) = -\cdot T\pi_i^n.$$

Moreover, the naturality square for $\eta\colon \mathrm{Id} \to Alg\, T \cdot F_T$ applied to π_i^n yields

$$\eta_n(i) = T\pi_i^n \quad \text{for all } i = 0, \ldots, n-1.$$

Let $\mathbb{M}$ be the monad corresponding to this adjoint situation. We know that $\mathbb{M}$ is finitary and that the values at $n \in \mathcal{N}$ are

$$Mn = \mathcal{T}(n, 1).$$

The theory $(Set_{\mathbb{M}}^f)^{op}$ thus has as morphisms from n to k precisely all functions

$$p\colon k \to \mathcal{T}(n, 1).$$

This k-tuple of morphisms defines a unique morphism

$$Ip \in \mathcal{T}(n, k)$$

characterized by

$$T\pi_i^k \cdot Ip = p(i) \quad (i = 0, \ldots, k-1).$$

We obtain an isomorphism of categories

$$I\colon (Set_{\mathbb{M}}^f)^{op} \to \mathcal{T},$$

and it remains to prove that the triangle

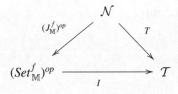

commutes. Given a morphism $u \in \mathcal{T}(n, k)$, we know that Tu is the unique morphism with

$$T\pi_i^k \cdot Tu = T\pi_{u(i)}^n$$

for all i, and we have

$$T\pi_i^n \cdot (I \cdot (J_\mathbb{M}^f)^{op}(u)) = T(\pi_i^k)I(\eta_n \cdot u) = \eta_n \cdot u(i) = \pi_{u(i)}^n.$$

□

A.38 Remark Explicitly, the functor assigning to every one-sorted algebraic theory $(\mathcal{T}, T)$ its monad of free $\mathcal{T}$-algebras is an equivalence functor from Th^1 to *FMon*. In fact, this functor is a quasi-inverse (0.2) of the equivalence functor E of the preceding proof.

A.39 Remark The situation with S-sorted theories and S-sorted algebraic categories is entirely analogous: all the preceding results translate without problems from *Set* to *Set*^{S}; only Theorem A.21 needs some work.

A.40 Theorem *S-sorted equational categories are up to concrete isomorphism precisely the categories $(Set^S)^\mathbb{M}$ of Eilenberg–Moore algebras for finitary monads $\mathbb{M}$ on Set^S.*

Proof The only difference with respect to the proof of A.21 is that in the S-sorted case, we have to check also that $(Set^S)^\mathbb{M}$ is closed in M-*Alg* under directed unions. Let $k_i: (X_i, x_i) \to (X, x)$ $(i \in I)$ be a colimit cocone of a filtered colimit in M-*Alg*, where each (X_i, x_i) is an Eilenberg–Moore algebra. Then Triangle (A.6) commutes for (X, x) because the cocone (k_i) is collectively epimorphic:

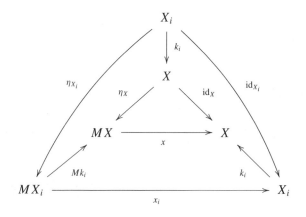

and Square (A.7) also commutes because the cocone (MMk_i) is collectively epimorphic, being a colimit cocone (since MM preserves filtered colimits):

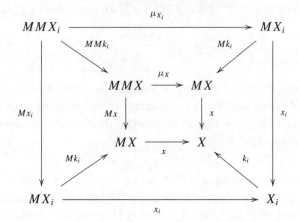

$\square$

A.41 Theorem *The category Th^S of S-sorted algebraic theories is equivalent to the category of finitary monads on Set^S.*

The proof is completely analogous to the proof in the one-sorted case. We just observe that the functor $J: Set^S \to (Set^S)^{\mathbb{M}}$ can, in case of M finitary, be restricted to $J^f: (S^*)^{op} \to (Set^S)^f_{\mathbb{M}}$ analogously to A.35.

Appendix B
Abelian categories

Another important topic not treated in our book is abelian categories. In this appendix, we restrict ourselves to introducing the basic concepts and proving that the only one-sorted algebraic categories that are abelian are the categories *R-Mod* of left modules over rings R. We also prove the many-sorted generalization of this result.

B.1 Remark In the following, we use the standard terminology of the theory of abelian categories:

1. A *zero object* is an object 0 that is initial as well terminal. For two objects A, B, the composite $A \to 0 \to B$ is denoted by 0: $A \to B$.
2. A *biproduct* of objects A and B is a product $A \times B$ with the property that the morphisms
   $$\langle \mathrm{id}_A, 0\rangle \colon A \to A \times B \quad \text{and} \quad \langle 0, \mathrm{id}_B\rangle \colon B \to A \times B$$
   form a coproduct of A and B.
3. A category is called *preadditive* if it is enriched over the category *Ab* of abelian groups, that is, if every hom-set carries the structure of an abelian group such that composition is a group homomorphism.
4. In a preadditive category, an object is a zero object iff it is terminal, and a product of two objects is a biproduct. A preadditive category with finite products is called *additive*.
5. A functor $F \colon \mathcal{A} \to \mathcal{A}'$ between preadditive categories is called *additive* if it is enriched over *Ab*, that is, the derived functions $\mathcal{A}(A, B) \to \mathcal{A}'(FA, FB)$ are group homomorphisms. In case of additive categories, this is equivalent to F preserving finite products.
6. Finally, a category is called *abelian* if it is exact and additive.

B.2 Example Just as one-object categories are precisely the monoids, one-object preadditive categories are precisely the unitary rings R. Every left R-module M defines an additive functor $\overline{M}: R \to Ab$ with $\overline{M}(*) = M$ and $\overline{M}r = r \cdot -: M \to M$ for $r \in R$. Conversely, every additive functor $F: R \to Ab$ is naturally isomorphic to $\overline{M}$ for $M = F(*)$.

For a small, preadditive category $\mathcal{C}$, we denote by $Add\,[\mathcal{C}, Ab]$ the abelian category of all additive functors into Ab (and all natural transformations). The previous example implies that $R\text{-}Mod$ is equivalent to $Add\,[R, Ab]$.

B.3 Theorem *Abelian algebraic categories are precisely those equivalent to*

$$Add\,[\mathcal{C}, Ab]$$

for a small additive category $\mathcal{C}$.

Proof: Sufficiency. For every small additive category $\mathcal{C}$, we prove that the category $Add\,[\mathcal{C}, Ab]$ is equivalent to $Alg\,\mathcal{C}$. We denote by $\hom(\mathcal{C}, -): \mathcal{C} \to Set$ and $\mathrm{Hom}(\mathcal{C}, -): \mathcal{C} \to Ab$ the hom-functors. Consider the forgetful functor $U: Ab \to Set$. Since U preserves finite products, it induces a functor

$$\widehat{U} = U \cdot -: Add\,[\mathcal{C}, Ab] \to Alg\,\mathcal{C}.$$

Let us prove that $\widehat{U}$ is an equivalence functor.

1. $\widehat{U}$ is faithful. This is obvious because U is faithful.
2. $\widehat{U}$ is full. In fact, we first observe that $\widehat{U}$ preserves sifted colimits. This follows from the fact that sifted colimits commute in Ab (as in any algebraic category; see 2.7) with finite products, and the functor $U = \hom(\mathbb{Z}, -)$ preserves sifted colimits. Next we verify that $\widehat{U}$ is full for morphisms (natural transformations)

$$\alpha: \widehat{U}(\mathrm{Hom}(C, -)) \to \widehat{U}G,$$

where $C \in \mathrm{obj}\,\mathcal{C}$ and $G: \mathcal{C} \to Ab$ is additive. By the Yoneda lemma, for all $X \in \mathcal{C}$ and for all $x: C \to X$, we have $\alpha_X(x) = Gx(a)$, where $a = \alpha_C(\mathrm{id}_C)$; thus α_X is a group homomorphism. Consequently, α lies in the image of $\widehat{U}$.

The general case of a morphism $\beta: \widehat{U}F \to \widehat{U}G$, where $F: \mathcal{C} \to Ab$ is additive, reduces to the previous case by using the fact that F is a sifted colimit of enriched representables and that $\widehat{U}$ preserves sifted colimits. To prove that F is a sifted colimit of representables, observe that, following 4.2, $\widehat{U}F$ is one. Now

$$\widehat{U}F = colim\,\hom(C_i, -) = colim\,(U \cdot \mathrm{Hom}(C_i, -))$$
$$= colim\,\widehat{U}(\mathrm{Hom}(C_i, -)) = \widehat{U}(colim\,\mathrm{Hom}(C_i, -)).$$

This implies that $F = \mathrm{colim}\,\mathrm{Hom}(C_i, -)$ because $\widehat{U}$ reflects sifted colimits (since it preserves sifted colimits and reflects isomorphisms).

3. $\widehat{U}$ is essentially surjective. We first take the representable functors $\hom(C, -)$ for $C \in \mathrm{obj}\,\mathcal{C}$, which are objects of $Alg\,\mathcal{C}$. Since $\mathcal{C}$ is preadditive, for every object C', the hom-set $\hom(C, C')$ is an abelian group, and $\hom(C, -)$ factorizes as

$$\mathcal{C} \xrightarrow{\mathrm{Hom}(C,-)} Ab \xrightarrow{U} Set,$$

with $\mathrm{Hom}(C, -)\colon \mathcal{C} \to Ab$ being additive. Therefore $\widehat{U}(\mathrm{Hom}(C, -)) \simeq \hom(C, -)$. Once again, the general case follows from the previous case by using the fact that any $\mathcal{C}$-algebra is a sifted colimit of representable $\mathcal{C}$-algebras and that $\widehat{U}$ preserves sifted colimits.

Necessity. Let $\mathcal{T}$ be an algebraic theory, and assume that $Alg\,\mathcal{T}$ is abelian. Since $\mathcal{T}^{op}$ embeds into $Alg\,\mathcal{T}$, $\mathcal{T}$ is preadditive (with finite products), and then it is a small additive category. Following the first part of the proof, $Alg\,\mathcal{T}$ is equivalent to $Add\,[\mathcal{T}, Ab]$. □

B.4 Corollary *Abelian algebraic categories are precisely the additive cocomplete categories with a strong generator consisting of perfectly presentable objects.*

In fact, this follows from 6.9 and B.3.

B.5 Remark In B.3, the condition that $\mathcal{C}$ is additive can be weakened: preadditivity is enough. In fact, let $\mathcal{C}$ be a small preadditive category. We can construct the small and preadditive category $Mat\,(\mathcal{C})$ of matrices over $\mathcal{C}$ as follows:

Objects are finite (possibly empty) families $(X_i)_{i \in I}$ of objects of $\mathcal{C}$.
Morphisms from $(X_i)_{i \in I}$ to $(Z_j)_{j \in J}$ are matrices $M = (m_{i,j})_{(i,j) \in I \times J}$ of morphisms $m_{i,j}\colon X_i \to Z_j$ in $\mathcal{C}$.
The matrix multiplication, the identity matrices, and matrix addition, as well known from linear algebra, define the composition, the identity morphisms, and the preadditive structure, respectively.

This new category $Mat\,(\mathcal{C})$ is additive. Indeed, it has a zero object given by the empty family and biproducts $\oplus$ given by disjoint unions. Let us check that the obvious embedding $\mathcal{C} \to Mat\,(\mathcal{C})$ induces an equivalence between $Add\,[Mat\,(\mathcal{C}), Ab]$ and $Add\,[\mathcal{C}, Ab]$. Indeed, given $F \in Add\,[\mathcal{C}, Ab]$, we get an extension $F' \in Add\,[Mat\,(\mathcal{C}), Ab]$ in the following way: $F'(M)$ is the unique

morphism such that the square

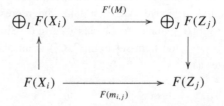

commutes for all $(i, j) \in I \times J$, where the vertical morphisms are injections in the coproduct and projections from the product, respectively. It is easy to verify that the functor $F \mapsto F'$ is an equivalence from $Add\,[\mathcal{C}, Ab]$ to $Add\,[Mat\,(\mathcal{C}), Ab]$.

B.6 Remark Observe that an object G of an additive, cocomplete category $\mathcal{A}$ is perfectly presentable iff its enriched hom-functor $\mathrm{Hom}(G, -)\colon \mathcal{A} \to Ab$ preserves colimits. (Compare with the absolutely presentable objects of 5.8.) In fact, if G is perfectly presentable, then $\mathrm{Hom}(G, -)$ preserves finite coproducts (because they are finite products) and reflexive coequalizers (because $U\colon Ab \to Set$ reflects them). This implies that $\mathrm{Hom}(G, -)$ preserves finite colimits. Indeed, given a parallel pair $a, b\colon X \rightrightarrows Z$ in $\mathcal{A}$, its coequalizer is precisely the coequalizer of the reflexive pair $(a, \mathrm{id}_Z), (b, \mathrm{id}_Z)\colon X + Z \rightrightarrows Z$. Finally, $\mathrm{Hom}(G, -)$ preserves arbitrary colimits because they are filtered colimits of finite colimits.

B.7 Example The group $\mathbb{Z}$ is perfectly presentable in Ab. Indeed,

$$\mathrm{Hom}(\mathbb{Z}, -)\colon Ab \to Ab$$

is naturally isomorphic to the identity functor. Observe that $\mathbb{Z}$ is, of course, not absolutely presentable.

B.8 Corollary *One-sorted abelian algebraic categories are precisely the categories equivalent to R-Mod for a unitary ring R.*

Proof Following B.3, a one-sorted abelian algebraic category $\mathcal{A}$ is of the form $Add\,[\mathcal{T}, Ab]$ for $\mathcal{T}$ a one-sorted additive algebraic theory with objects T^n ($n \in \mathbb{N}$). Any $F \in Add\,[\mathcal{T}, Ab]$ can be restricted to an additive functor $\mathcal{T}(T, T) \to Ab$, where the ring $\mathcal{T}(T, T)$ is seen as a preadditive category with a single object. Moreover, F is uniquely determined by such a restriction because each object of $\mathcal{T}$ is a finite product of T. Finally, $Add\,[\mathcal{T}(T, T), Ab]$ is equivalent to $\mathcal{T}(T, T)$-Mod. □

B.9 Corollary *Finitary localizations of abelian algebraic categories are precisely the abelian locally finitely presentable categories.*

Proof Let $\mathcal{A}$ be an abelian, locally finitely presentable category. Following the proof of 18.10, we have that $\mathcal{A} = Lex\,\mathcal{T}$ is a finitary localization of $Alg\,\mathcal{T}$, with $\mathcal{T}$ an additive algebraic theory. Because of B.3, $Alg\,\mathcal{T}$ is equivalent to $Add\,[\mathcal{T}, Ab]$. □

Appendix C
More about dualities for one-sorted algebraic categories

Throughout the book, we took the "strict view" of what a theory morphism or concrete functor or monadic functor should be; that is, the condition put on the functor in question was formulated as an equality between two functors. There is a completely natural nonstrict view wherein the conditions are formulated as natural isomorphisms between functors. This has a number of advantages. For example, we can present a characterization of one-sorted algebraic categories (see Theorem C.6) for which we know no analogous result in the strict variant. Also, the duality between one-sorted algebraic theories and uniquely transportable one-sorted algebraic categories can be directly derived from the nonstrict version of the biduality 11.38 without using monads. In this appendix, we briefly mention the nonstrict variants of some concepts in the book.

C.1 Definition

1. Given concrete categories $U: \mathcal{A} \to \mathcal{K}$ and $V: \mathcal{B} \to \mathcal{K}$ by a *pseudoconcrete functor* between them, we mean a functor $F: \mathcal{A} \to \mathcal{B}$ such that $V \cdot F$ is naturally isomorphic to U.
2. Given concrete categories $U: \mathcal{A} \to \mathcal{K}$ and $V: \mathcal{B} \to \mathcal{K}$ by a *pseudoconcrete equivalence* between them, we mean a functor $F: \mathcal{A} \to \mathcal{B}$, which is at the same time an equivalence and pseudoconcrete. (Note that any quasi-inverse of F is necessarily pseudoconcrete.) We then say that $(\mathcal{A}, U)$ and $(\mathcal{B}, V)$ are *pseudoconcretely equivalent*.

C.2 Definition A concrete category $(\mathcal{A}, U)$ on $\mathcal{K}$ is *pseudomonadic* if there exists a monad $\mathbb{M}$ on $\mathcal{K}$ and a pseudoconcrete equivalence $\mathcal{A} \to \mathcal{K}^{\mathbb{M}}$.

C.3 Beck's theorem: *Characterization of pseudomonadic categories A concrete category $(\mathcal{A}, U)$, with U a right adjoint, is pseudomonadic iff (1) U is*

More about dualities for one-sorted algebraic categories 233

conservative, and (2) $\mathcal{A}$ has coequalizers of all reflexive pairs f, g such that Uf, Ug have an absolute coequalizer; and U preserves these coequalizers.

A proof of C.3 can be found in Borceux (1994, chapt. 4, sect. 4). The reader just needs to observe that the parallel pairs of morphisms used in that proof are reflexive.

C.4 Proposition *For every one-sorted algebraic theory* $(\mathcal{T}, T)$, *the concrete category* $(Alg\,\mathcal{T}, Alg\,T)$ *is pseudomonadic.*

Proof The functor $Alg\,T$ is algebraic and conservative (11.8), and algebraic functors are right adjoints and preserve reflexive coequalizers. Following C.3, $(Alg\,\mathcal{T}, Alg\,T)$ is pseudomonadic. □

C.5 Definition A *pseudo-one-sorted algebraic category* is a concrete category over *Set* that is pseudoconcretely equivalent to $Alg\,T: Alg\,\mathcal{T} \to Set$ for a one-sorted algebraic theory $(\mathcal{T}, T)$.

C.6 Theorem: *Characterization of pseudo-one-sorted algebraic categories*
The following conditions on a concrete category $(\mathcal{A}, U)$ *over Set are equivalent:*

1. $(\mathcal{A}, U)$ *is pseudo-one-sorted algebraic.*
2. $\mathcal{A}$ *is cocomplete, and U is a conservative right adjoint preserving sifted colimits.*

In more detail, let $\mathcal{A}$ be a cocomplete category. Given a faithful functor

$$U: \mathcal{A} \to Set$$

with $\mathcal{A}$ cocomplete, there exists a one-sorted algebraic theory $(\mathcal{T}, T)$ and an equivalence functor

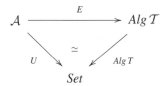

making the preceding triangle commutative up to natural isomorphism iff U is a conservative right adjoint preserving sifted colimits.

Proof The conditions are necessary by 11.8. Let us prove that they are sufficient: let $U: \mathcal{A} \to Set$ be as in item 2, with F a left adjoint of U. The set $\mathcal{F} = \{Fn : n \in \mathbb{N}\}$ is closed in $\mathcal{A}$ under finite coproducts, and by 6.16, it is a strong generator formed by perfectly presentable objects. Following the proof

of 6.9 (implication 3 $\Rightarrow$ 1), the functor
$$E: \mathcal{A} \to Alg(\mathcal{F}^{op}), \quad K \mapsto \mathcal{A}(-, K): \mathcal{F}^{op} \to Set$$
is an equivalence. Moreover, the codomain restriction $T: \mathcal{N} \to \mathcal{F}^{op}$ of the functor $F^{op} \cdot Y_{\mathcal{N}}: \mathcal{N} \to \mathcal{A}^{op}$ is a morphism of theories. Let us check that there exists a natural isomorphism $Alg\, T \cdot E \simeq U$: for every $A \in \mathcal{A}$ and $n \in \mathbb{N}$, the functor $Alg\, T \cdot E(A)$ assigns to $n \in \mathbb{N}$ the set $\mathcal{A}(Tn, A)$, and since $Tn = F(Y_{\mathcal{N}}(n))$, the adjunction $F \dashv U$ clearly yields a natural isomorphism $\mathcal{A}(Tn, A) \simeq UA(n)$.

Factorize T as a functor $T': \mathcal{N} \to \mathcal{T}$, which is the identity on objects and equal to T on morphisms, followed by a functor $T'': \mathcal{T} \to \mathcal{F}^{op}$, which is the identity on morphisms and equals to T on objects. Therefore T'' is an equivalence (because T is surjective on objects), and $(\mathcal{T}, T')$ is a one-sorted theory. The following diagram concludes the proof:

$$\begin{array}{ccccc}
\mathcal{A} & \xrightarrow{E} & Alg(\mathcal{F}^{op}) & \xrightarrow{Alg\, T''} & Alg\, \mathcal{T} \\
 & \searrow{\simeq} & \downarrow{Alg\, T} & \swarrow{Alg\, T'} & \\
 & U & Set & &
\end{array}$$

$\square$

C.7 Corollary *Pseudo-one-sorted algebraic categories are up to pseudoconcrete equivalence precisely the categories $Set^{\mathbb{M}}$ of Eilenberg–Moore algebras for finitary monads $\mathbb{M}$ on Set.*

In fact, this follows from C.4, C.6, and A.23.

It is easy to extend the biequivalence of Theorem 11.38 to pseudo-one-sorted algebraic categories.

C.8 Definition Given one-sorted algebraic theories $(\mathcal{T}_1, T_1)$ and $(\mathcal{T}_2, T_2)$, a *pseudomorphism*
$$M: (\mathcal{T}_1, T_1) \to (\mathcal{T}_2, T_2)$$
is a functor $M: \mathcal{T}_1 \to \mathcal{T}_2$, with $M \cdot T_1$ naturally isomorphic to T_2.

C.9 Remark Pseudomorphisms preserve finite products.

C.10 Theorem: *Nonstrict one-sorted algebraic duality* *The 2-category of*

objects: one-sorted algebraic theories,
1-cells: pseudomorphisms
2-cells: natural transformations

is dually biequivalent to the 2-category of

objects: pseudo-one-sorted algebraic categories
1-cells: pseudoconcrete functors
2-cells: natural transformations

Proof The proof is analogous to the proof of 11.38; just observe that from a pseudoconcrete functor

$$(G, \varphi): (Alg\ T_2, Alg\ T_2) \to (Alg\ T_1, Alg\ T_1),$$

we get a natural isomorphism $\psi: F_{T_1} \to F \cdot F_{T_2}$ between the left adjoints. The rest of the proof of 11.38 can now be repeated with no other changes. □

C.11 Remark In Appendix A, we obtained the duality between the category of one-sorted algebraic theories and the category of uniquely transportable one-sorted algebraic categories, using finitary monads. We are going to derive such a duality from the biequivalence of Theorem C.10. To perform this, we need several preliminary steps. We start with an observation explaining why, when we restrict our attention to uniquely transportable one-sorted algebraic categories, we do not need a nonstrict version.

C.12 Lemma *Let* $(\mathcal{A}_1, U_1)$, $(\mathcal{A}_2, U_2)$ *be concrete categories over* $\mathcal{K}$.

1. *If* $(\mathcal{A}_2, U_2)$ *is transportable, then for every pseudoconcrete functor*

$$G: (\mathcal{A}_1, U_1) \to (\mathcal{A}_2, U_2),$$

   there exists a concrete functor $H: (\mathcal{A}_1, U_1) \to (\mathcal{A}_2, U_2)$ *naturally isomorphic to* G.

2. *If* $(\mathcal{A}_1, U_1)$ *and* $(\mathcal{A}_2, U_2)$ *are transportable, then they are pseudoconcretely equivalent iff they are concretely equivalent.*

Proof

1. Let $\varphi: U_1 \to U_2 \cdot G$ be a natural isomorphism. For every object $A \in \mathcal{A}_1$, consider the isomorphism

$$\varphi_A: U_1 A \to U_2 G A.$$

Since $(\mathcal{A}_2, U_2)$ is transportable, there exists an object HA in $\mathcal{A}_2$ and an isomorphism

$$\psi_A: HA \to GA$$

such that $U_2 \psi_A = \varphi_A$. This gives a map on objects $H: \mathcal{A}_1 \to \mathcal{A}_2$. For $f: A \to B$ in $\mathcal{A}_1$, put $Hf = \psi_B^{-1} \cdot Gf \cdot \psi_A$. In this way, H is a functor such that $U_2 \cdot H = U_1$, and $\psi: H \to G$ is a natural isomorphism.

2. This follows immediately from 1. □

C.13 Corollary *Every transportable pseudo-one-sorted algebraic category is one-sorted algebraic.*

Recall from 11.35 the 2-categories Th^1 of one-sorted algebraic theories and ALG^1 of one-sorted algebraic categories. In the remaining part of this appendix, we consider Th^1 and ALG^1 as categories; that is, we forget the 2-cells:

C.14 Definition

1. The category Th^1 of one-sorted theories has
   objects: one-sorted algebraic theories
   morphisms: morphisms of one-sorted algebraic theories
2. The category ALG^1 of one-sorted algebraic categories has
   objects: one-sorted algebraic categories
   morphisms: concrete functors
3. The category ALG^1_u is the full subcategory of ALG^1 of all uniquely transportable one-sorted algebraic categories.

C.15 Lemma *The category Th^1 (seen as a 2-category with only identity 2-cells) is biequivalent to the 2-category $PsTh^1$ having*

> *objects: one-sorted algebraic theories $(\mathcal{T}, T)$*
> *1-cells from $(\mathcal{T}_1, T_1)$ to $(\mathcal{T}_2, T_2)$: pairs (M, μ) where $M: \mathcal{T}_1 \to \mathcal{T}_2$ is a pseudomorphism of one-sorted theories and $\mu: M \cdot T_1 \to T_2$ is a natural isomorphism*
> *2-cells from (M, μ) to (N, ν): natural transformations $\alpha: M \to N$ that are coherent, that is, such that $\nu \cdot \alpha T_1 = \mu$*

Proof

1. Let us start by observing that the coherence condition $\nu \cdot \alpha T_1 = \mu$ on a 2-cell α of $PsTh^1$ immediately implies that α is invertible and that between two parallel 1-cells of $PsTh^1$ there is at most one 2-cell.

2. The inclusion $Th^1 \to PsTh^1$ is a biequivalence: since Th^1 and $PsTh^1$ have the same objects, we have to prove that the induced functor

$$Th^1((\mathcal{T}_1, T_1), (\mathcal{T}_2, T_2)) \to PsTh^1((\mathcal{T}_1, T_1), (\mathcal{T}_2, T_2))$$

is full and essentially surjective (it is certainly faithful because Th^1 has only identity 2-cells).

2a. Full: let $M, N: (\mathcal{T}_1, T_1) \to (\mathcal{T}_2, T_2)$ be 1-cells in Th^1 and $\alpha: (M, =) \to (N, =)$ be a 2-cell in $PsTh^1$. The coherence condition gives $\alpha_n = \text{id}$ for every $n \in \mathcal{N}$, and then $M = N$ by naturality of α.

2b. Essentially surjective: consider a 1-cell $(M, \mu): (\mathcal{T}_1, T_1) \to (\mathcal{T}_2, T_2)$ in $PsTh^1$. We define a functor

$$N: \mathcal{T}_1 \to \mathcal{T}_2, \quad N(f: x \to y) = \mu_y \cdot Mf \cdot \mu_x^{-1}: x \to y.$$

It is easy to check that $N \cdot T_1 = T_2$ and that $\mu: (M, \mu) \to (N, =)$ is a 2-cell in $PsTh^1$. □

C.16 Lemma *The category* ALG_u^1 *(seen as a 2-category with only identity 2-cells) is biequivalent to the 2-category* $PsALG^1$ *having*

objects: pseudo-one-sorted algebraic categories $(\mathcal{A}, U)$
1-cells from $(\mathcal{A}_1, U_1)$ *to* $(\mathcal{A}_2, U_2)$: *pairs* (G, φ) *where* $G: \mathcal{A}_1 \to \mathcal{A}_2$ *is a functor and* $\varphi: U_1 \to U_2 \cdot G$ *is a natural isomorphism*
2-cells from (G, φ) *to* (H, ψ): *natural transformations* $\alpha: G \to H$ *that are coherent, that is, such that* $U_2\alpha \cdot \varphi = \psi$

Proof

1. Let us start by observing that since U_2 is conservative, the coherence condition $U_2\alpha \cdot \varphi = \psi$ implies that α is invertible, and since U_2 is faithful, it implies that between two parallel 1-cells of $PsALG^1$, there is at most one 2-cell.

2. The inclusion $ALG_u^1 \to PsALG^1$ is essentially surjective (in the sense of the 2-category $PsALG^1$): let $(\mathcal{A}, U)$ be an object in $PsALG^1$. We are going to construct the diagram

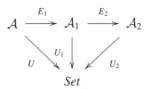

where E_1 and E_2 are pseudoconcrete equivalences, $(\mathcal{A}_1, U_1)$ is transportable, and $(\mathcal{A}_2, U_2)$ is uniquely transportable. Therefore $(\mathcal{A}_2, U_2)$ is a uniquely transportable pseudo-one-sorted algebraic category. By Corollary C.13, we conclude that $(\mathcal{A}_2, U_2)$ is an object of ALG_u^1.

2a. Objects in $\mathcal{A}_1$ are triples $(A, \pi_{A,X}: UA \to X, X)$, with $A \in \mathcal{A}$, X being a set and $\pi_{A,X}$ an isomorphism. A morphism from $(A, \pi_{A,X}, X)$ to $(A', \pi_{A',X'}, X')$ is a pair of morphisms $a: A \to A', x: X \to X'$ such that $x \cdot \pi_{A,X} = \pi_{A',X'} \cdot Ua$. Clearly the forgetful functor

$$U_1: \mathcal{A}_1 \to Set, \quad U_1(A, \pi_{A,X}, X) = X$$

is transportable. Moreover, we have an equivalence

$$E_1 \colon \mathcal{A} \to \mathcal{A}_1, \quad E_1 A = (A, \mathrm{id}_{UA}, UA),$$

with the quasi-inverse given by forgetful functor

$$E_1' \colon \mathcal{A}_1 \to \mathcal{A}, \quad E_1'(A, \pi_{A,X}, X) = A.$$

Note that E_1 is a concrete functor, whereas E_1' is pseudoconcrete: a natural isomorphism $\pi \colon U \cdot E_1' \to U_1$ is given by

$$\pi_{(A, \pi_{A,X}, X)} = \pi_{A,X}.$$

2b. Consider any concrete category $U_1 \colon \mathcal{A}_1 \to \mathit{Set}$. We define a category $\mathcal{A}_2$ whose objects are equivalence classes of objects of $\mathcal{A}_1$, with X being equivalent to X' if there exists a U_1-identity $i \colon X \to X'$, that is, an isomorphism i such that $U_1 i = \mathrm{id}_{U_1 X}$. We denote by $[X]$ the equivalence classe of X. The hom-set $\mathcal{A}_2([X][Z])$ is the quotient of the disjoint union of the $\mathcal{A}_1(X', Z')$ for $X' \in [X]$, $Z' \in [Z]$, with $f \colon X \to Z$ being equivalent to $f' \colon X' \to Z'$ if there exist U_1-identites $i \colon X \to X'$ and $j \colon Z \to Z'$ and $U_1 f = U_1 f'$. The composition of $[f] \colon [X] \to [Z]$ and $[g] \colon [Z] \to [W]$ is $[gjf] \colon [X] \to [W]$, where $j \colon Z \to Z'$ is any U_1-identity. The functors E_2 and U_2 are defined by

$$E_2 \colon \mathcal{A}_1 \to \mathcal{A}_2, \quad E_2(f \colon X \to Z) = [f] \colon [X] \to [Z],$$

$$U_2 \colon \mathcal{A}_2 \to \mathit{Set}, \quad U_2([f] \colon [X] \to [Z]) = Ff \colon FX \to FZ.$$

Clearly $U_2 \cdot E_2 = U_1$ and E_2 is full and surjective on objects, so that it is an equivalence (because U_1 is faithful). Finally, it is easy to check that U_2 is amnestic if U_1 is conservative and that U_2 is transportable if U_1 is transportable.

3. The induced functor

$$ALG_u^1((\mathcal{A}_1, U_1), (\mathcal{A}_2, U_2)) \to PsALG^1((\mathcal{A}_1, U_1), (\mathcal{A}_2, U_2))$$

is an equivalence.

3a. Full: let $G, H \colon (\mathcal{A}_1, U_1) \to (\mathcal{A}_2, U_2)$ be 1-cells in ALG_u^1 and $\alpha \colon (G, =) \to (H, =)$ a 2-cell in $PsALG^1$. The coherence condition gives $U_2(\alpha_A) = \mathrm{id}$ for every $A \in \mathcal{A}$. Since U_2 is amnestic, α_A is the identity.

3b. Faithful: this is obvious because ALG^1 has only identity 2-cells.

3c. Essentially surjective: let $(\mathcal{A}_1, U_1), (\mathcal{A}_2, U_2)$ be objects in ALG_u^1 and $(G, \varphi) \colon (\mathcal{A}_1, U_1) \to (\mathcal{A}_2, U_2)$ a 1-cell in $PsALG^1$. As in the proof of C.12, we get a concrete functor $H \colon (\mathcal{A}_1, U_1) \to (\mathcal{A}_2, U_2)$ and a natural isomorphism $\psi \colon H \to G$. To end the proof, observe that $\psi \colon (H, =) \to (G, \varphi)$ is a 2-cell in $PsALG^1$. Indeed the condition $U_2(\psi_A) = \varphi_A$ is precisely the coherence condition on ψ. $\square$

Recall the 2-fucntor $Alg^1: (Th^1)^{op} \to ALG^1$ from 11.37. There is an obvious version of this 2-functor in the present context: all we need to observe is that given a coherent natural transformation α in $PsTh^1$, it follows that $Alg\,\alpha$ is also coherent. By a slight abuse of notation, we denote this 2-functor by Alg^1 again:

C.17 Notation We denote by

$$Alg^1: (PsTh^1)^{op} \to PsALG^1$$

the 2-functor assigning to every one-sorted theory $(\mathcal{T}, T)$ the one-sorted algebraic category $(Alg\,\mathcal{T}, Alg\,T)$, to every 1-cell $(M, \mu): (\mathcal{T}_1, T_1) \to (\mathcal{T}_2, T_2)$ the 1-cell $(Alg\,M, Alg\,\mu^{-1})$, and to every 2-cell $\alpha: (M, \mu) \to (N, \nu)$ the 2-cell $Alg\,\alpha$ whose component at a $\mathcal{T}_2$-algebra A is $A \cdot \alpha: A \cdot M \to A \cdot N$.

C.18 Theorem: *One-sorted algebraic duality* *The category* ALG^1_u *of uniquely transportable one-sorted algebraic categories is equivalent to the dual of the category* Th^1 *of one-sorted algebraic theories. In fact, the 2-functor*

$$Alg^1: (PsTh^1)^{op} \to PsALG^1$$

is a biequivalence.

Proof

1. The 2-functor Alg^1 is well defined by 11.8.

2. Alg^1 is essentially surjective (in the sense of the 2-category $PsALG^1$): following C.6, for every object $(\mathcal{A}, U)$ of $PsALG^1$, there exists a pseudoconcrete equivalence $E: \mathcal{A} \to Alg\,\mathcal{T}$ with natural isomorphism

$$\varphi: Alg\,T \cdot E \to U.$$

Recall from 0.3 that it is possible to choose a quasi-inverse $E': Alg\,\mathcal{T} \to \mathcal{A}$ and natural isomorphisms

$$\eta: \mathrm{Id}_{Alg\,\mathcal{T}} \to E \cdot E' \quad \text{and} \quad \varepsilon: E' \cdot E \to \mathrm{Id}_\mathcal{A}$$

such that

$$E\varepsilon \cdot \eta E = E \quad \text{and} \quad \varepsilon E' \cdot E'\eta = E'.$$

We get a natural isomorphism

$$\psi = \varphi^{-1} E' \cdot Alg\,T\eta: Alg\,T \to U \cdot E'.$$

It follows that

$$\eta: (\mathrm{Id}_{Alg\,\mathcal{T}}, =) \to (E \cdot E', \varphi E' \cdot \psi) \quad \text{and} \quad \varepsilon: (E' \cdot E, \psi E \cdot \varphi) \to (\mathrm{Id}_\mathcal{A}, =)$$

are 2-cells in $PsALG^1$. Indeed, the coherence condition on η is just the definition of ψ, and the coherence condition on ε follows from the equation $E\varepsilon \cdot \eta E = E$:

$$U\varepsilon \cdot \psi E \cdot \varphi = U\varepsilon \cdot \varphi^{-1} E'E \cdot Alg\, T\eta E \cdot \varphi$$
$$= \varphi^{-1} \cdot Alg\, TE\varepsilon \cdot Alg\, T\eta E \cdot \varphi = \varphi^{-1} \cdot \varphi = U.$$

We conclude that $(\mathcal{A}, U)$ and $(Alg\, \mathcal{T}, Alg\, T)$ are equivalent objects in $PsALG^1$.

3. We will prove that for two one-sorted algebraic theories $(\mathcal{T}_1, T_1)$ and $(\mathcal{T}_2, T_2)$, the functor $Alg^1_{(\mathcal{T}_1, T_1), (\mathcal{T}_2, T_2)}$ from $PsTh^1((\mathcal{T}_1, T_1), (\mathcal{T}_2, T_2))$ to $PsALG^1((Alg\, \mathcal{T}_2, Alg\, T_2), (Alg\, \mathcal{T}_1, Alg\, T_1))$ is an equivalence of categories.

3a. Full and faithful: the proof is as in Theorem 9.15; indeed, $\alpha\colon M \to N$ is coherent iff $Alg\, \alpha\colon Alg\, M \to Alg\, N$ is coherent.

3b. Essentially surjective: we follow the proof of 11.38. Let

$$(G, \varphi)\colon (Alg\, \mathcal{T}_2, Alg\, T_2) \to (Alg\, \mathcal{T}_1, Alg\, T_1)$$

be a 1-cell in $PsALG^1$ and $\psi\colon F_{T_2} \to F \cdot F_{T_1}$ the induced natural isomorphism on the adjoint functors. As in 11.38, we get a 1-cell

$$(M, =)\colon (\mathcal{T}_1, T_1) \to (\mathcal{T}_2, T_2)$$

in $PsTh^1$, and we have to construct a 2-cell

$$\alpha\colon (Alg\, M, =) \to (G, \varphi)$$

in $PsALG^1$. Since $Alg\, T_1 \cdot Alg\, M = Alg\, T_2$, there exists a natural isomorphism $i\colon F_{T_2} \to M^* \cdot F_{T_1}$. Let

$$\psi_|\colon Y_{\mathcal{T}_2} \cdot M^{op} \to F \cdot Y_{\mathcal{T}_1}, \quad i_|\colon Y_{\mathcal{T}_2} \cdot M^{op} \to M^* \cdot Y_{\mathcal{T}_1}$$

be the restrictions of ψ and i to $\mathcal{T}_1$. By 9.3, there exists a natural isomorphism $\alpha^*\colon M^* \to F$. Moreover, α^* is unique with the condition $\alpha^* Y_{\mathcal{T}_1} \cdot i_| = \psi_|$ (apply 4.11 to $Y_{\mathcal{T}_1}$). Since F_{T_2} and $F \cdot F_{T_1}$ preserve colimits, the previous equation gives $\alpha^* F_{T_1} \cdot i = \psi$ (apply 4.11 to $Y_\mathcal{N}$). Passing to the right adjoints, we get a natural isomorphism $\alpha\colon Alg\, M \to G$ such that $(Alg\, T_1)\alpha = \varphi$, as required. $\square$

References

Adámek, J. (1974). Free algebras and automata realizations in the language of categories. *Comment. Math. Univ. Carolina.* **14**: 589–602.
Adámek, J. (1977). Colimits of algebras revisited. *Bull. Austral. Math. Soc.* **17**: 433–450.
Adámek, J., F. Borceux, S. Lack, and J. Rosický (2002). A classification of accessible categories. *J. Pure Appl. Algebra* **175**: 7–30.
Adámek, J., H. Herrlich, and G. Strecker (2009). *Abstract and Concrete Categories*. Dover.
Adámek, J., F. W. Lawvere, and J. Rosický (2001a). How algebraic is algebra? *Theory Appl. Categ.* **8**: 253–283.
Adámek, J., F. W. Lawvere, and J. Rosický (2003). On the duality between varieties and algebraic theories. *Alg. Univ.* **49**: 35–49.
Adámek, J., and H.-E. Porst (1998). Algebraic theories of quasivarieties. *J. Algebra* **208**: 379–398.
Adámek, J., and J. Rosický (1994). *Locally Presentable and Accessible Categories*. Cambridge University Press.
Adámek, J., and J. Rosický (2001). On sifted colimits and generalized varieties. *Theory Appl. Categ.* **8**: 33–53.
Adámek, J., J. Rosický, and E. M. Vitale (2001b). On algebraically exact categories and essential localizations of varieties. *J. Algebra* **244**: 450–477.
Adámek, J., J. Rosický, and E. M. Vitale (2010). What are sifted colimits? *Theory Appl. Categ.* **23**: 251–260
Adámek, J., M. Sobral, and L. Sousa (2006). Morita equivalence for many-sorted algebraic theories. *J. Algebra* **297**: 361–371.
Almeida, J. (1994). *Finite Semigroups and Universal Algebra*. World Scientific.
Artin, M., A. Grothendieck, and J. L. Verdier (1972). *Théorie des topos et cohomologie étale des schémas*. Lect. Notes in Math. **269**. Springer.
Banaschewski, B. (1972). Functors into categories of M-sets. *Abh. Mat. Sem. Univ. Hamburg* **38**: 49–64.
Barr, M. (1970). Coequalizers and free triples. *Math. Z.* **116**: 307–322.
Barr, M., A. P. Grillet, and D. H. van Osdol (1971). *Exact Categories and Categories of Sheaves*. Lect. Notes in Math. **236**. Springer.
Barr, M., and C. Wells (1985). *Toposes, Triples and Theories*. Springer.
Barr, M., and C. Wells (1990). *Category Theory for Computing Science*. Prentice Hall.

Bass, H. (1968). *Algebraic K-Theory*. Benjamin.
Bastiani, A., and C. Ehresmann (1972). Categories of sketched structures. *Cah. Top. Géom. Différ. Catég.* **13**: 104–214.
Bénabou, J. (1967). Introduction to bicategories. In *Reports of the Midwest Category Seminar*. Lect. Notes in Math. **40**. Springer, 1–77.
Bénabou, J. (1968). Structures algébriques dans les catégories. *Cah. Top. Géom. Différ. Catég.* **10**: 1–126.
Birkhoff, G. (1935). On the structure of abstract algebras. *Proc. Cambr. Phil. Soc.* **31**: 433–454.
Birkhoff, G., and J. D. Lipson (1970). Heterogenous algebras. *J. Combin. Theory* **8**: 115–133.
Boardman, J. M., and R. M. Vogt (1973). *Homotopy Invariant Algebraic Structures on Topological Spaces*. Lect. Notes in Math. **347**. Springer.
Borceux, F. (1994). *Handbook of Categorical Algebra*. Cambridge University Press.
Borceux, F., and E. M. Vitale (1994). On the notion of bimodel for functorial semantics. *Appl. Categ. Struct.* **2**: 283–295.
Bourbaki, N. (1956). *Théorie des Ensembles*. Herrman.
Bunge, M. (1966). Categories of set-valued functors. Dissertation, University of Pennsylvania.
Bunge, M., and A. Carboni (1995). The symmetric topos. *J. Pure Appl. Algebra* **105**: 233–249.
Carboni, A., and R. Celia Magno (1982). The free exact category on a left exact one. *J. Austral. Math. Soc. Ser. A* **33**: 295–301.
Carboni, A., and E. M. Vitale (1998). Regular and exact completions. *J. Pure Appl. Algebra* **125**: 79–116.
Centazzo, C., J. Rosický, and E. M. Vitale (2004). A characterization of locally D-presentable categories. *Cah. Top. Géom. Différ. Catég.* **45**: 141–146.
Centazzo, C., and E. M. Vitale (2002). A duality relative to a limit doctrine. *Theory Appl. Categ.* **10**: 486–497.
Cohn, P. M. (1965). *Universal Algebra*. Harper and Row.
Diers, Y. (1976). Type de densité d'une sous-catégorie pleine. *Ann. Soc. Sc. Bruxelles* **90**: 25–47.
Dukarm, J. J. (1988). Morita equivalence of algebraic theories. *Colloq. Math.* **55**: 11–17.
Ehresmann, C. (1963). Catégories structurées. *Ann. Sci. École Norm. Sup.* **80**: 349–426.
Ehresmann, C. (1967). Sur les structures algébriques. *C. R. Acad. Sci. Paris* **264**: A840–A843.
Eilenberg, S. (1961). Abstract description of some basic functors. *J. Indian Math. Soc. (N. S.)* **24**: 231–234.
Elkins, B., and J. A. Zilber (1976). Categories of actions and Morita equivalence. *Rocky Mountain J. Math.* **6**: 199–225.
Gabriel, P. (1962). Des catégories abéliennes. *Bull. Sci. Math. France* **90**: 323–448.
Gabriel, P., and F. Ulmer (1971). *Lokal Präsentierbare Kategorien*. Lect. Notes in Math. **221**. Springer.
Gran, M., and E. M. Vitale (1998). On the exact completion of the homotopy category. *Cahiers Top. Géom. Diff. Cat.* **39**: 287–297.
Grätzer, G. (2008). *Universal Algebra*. Springer.

Hagemann, J., and C. Hermann (1979). A concrete ideal multiplication for algebraic systems and its relation to congruence modularity. *Arch. Math. (Basel)* **32**: 234–245.

Heller, A. (1965). MR0163940. *Math. Rev.* **29**.

Higgins, P. J. (1963–1964). Algebras with a scheme of operators. *Math. Nachr.* **27**: 115–132.

Hyland, M., and J. Power (2007). The category theoretic understanding of universal algebra: Lawvere theories and monads. *Electron. Notes Theor. Comput. Sci.* **172**: 437–458.

Janelidze, Z. (2006). Closedness properties of internal relations I: A unified approach to Mal'tsev, unital and subtractive categories. *Theory Appl. Categ.* **16**: 236–261.

Janelidze, G., and M. C. Pedicchio (2001). Pseudogroupoids and commutators. *Theory Appl. Categ.* **8**: 408–456.

Joyal, A. (2008). Notes on logoi (unpublished book in preparation).

Kelly, G. M. (1982). *Basic Concepts of Enriched Categories*. Cambridge University Press.

Lack, S., and J. Rosický (2010). Notions of Lawvere theory. *Appl. Cat. Struct.* (In press).

Lair, C. (1996). Sur le genre d'esquissabilité des catégories modelables (accessibles) possédant les produits de deux. *Diagrammes* **35**: 25–52.

Lambek, J. (1968). A fixpoint theorem for complete categories. *Math. Z.* **103**: 151–161.

Lawvere, F. W. (1963). Functorial semantics of algebraic theories. Dissertation, Columbia University.

Lawvere, F. W. (1969). Ordinal sums and equational doctrines. In *Seminar on Triples and Categorical Homology Theory*. Lect. Notes in Math. **80**: Springer, 141–155.

Lawvere, F. W. (2008). Core varieties, extensivity, and rig geometry. *Theory Appl. Categ.* **20**: 497–503.

Linton, F. E. J. (1966). Some aspects of equational categories. In *Proc. Conf. Categorical Algebra La Jolla 1965*. Springer, 84–94.

Linton, F. E. J. (1969a). Coequalizers in categories of algebras. In *Seminar on Triples and Categorical Homology Theory*. Lect. Notes in Math. **80**. Springer, 75–90.

Linton, F. E. J. (1969b). Applied functorial semantics, II. In *Seminar on Triples and Categorical Homology Theory*. Lect. Notes in Math. **80**. Springer, 53–74.

Loday, J.-L. (2008). *Generalized Bialgebras and Triples of Operads*. Astérisque.

Lurie, J. (2009). *Higher Topos Theory*. Princeton University Press.

Mac Lane, S. (1963). Natural associativity and commutativity. *Rice Univ. Stud.* **49**: 38–46.

Mac Lane, S. (1998). *Categories for the Working Mathematician*. Springer.

Makkai, M., and R. Paré (1989). *Accessible Categories: The Foundation of Categorical Model Theory*. Contemp. Math. **104**. AMS.

Manes, E. G. (1976). *Algebraic Theories*. Springer.

McKenzie, R. N. (1996). An algebraic version of categorical equivalence for varieties and more general categories. Lect. Notes Pure Appl. Mathematics **180**. Marcel Dekker, 211–243.

Mitchell, B. (1965). *Theory of Categories*. Academic.

Morita, K. (1958). Duality for modules and its applications to the theory of rings with minimum conditions. *Sci. Rep. Tokyo Kyoiku Daigaku, Sect. A* **6**: 83–142.

Pareigis, B. (1970). *Categories and Functors*. Academic.

Pedicchio, M. C. (1995). A categorical approach to commutator theory. *J. Algebra* **177**: 647–657.

Pedicchio, M. C., and J. Rosický (1999). Comparing coequalizer and exact completions. *Theory Appl. Categ.* **6**: 77–82.

Pedicchio, M. C., and F. Rovatti (2004). Algebraic categories. In *Categorical Foundations*. Cambridge University Press, 269–310.

Pedicchio, M. C., and E. M. Vitale (2000). On the abstract characterization of quasivarieties. *Algebra Universalis* **43**: 269–278.

Pedicchio, M. C., and R. J. Wood (2000). A simple characterization of theories of varieties. *J. Algebra* **233**: 483–501.

Popesco, N., and P. Gabriel (1964). Caractérisation des catégories abéliennes avec générateurs et limites inductives exactes. *C. R. Acad. Sci. Paris* **258**: 4188–4190.

Porst, H.-E. (2000). Equivalence for varieties in general and for Bool in particular. *Algebra Universalis* **43**: 157–186.

Power, J. (1999). Enriched Lawvere theories. *Theory Appl. Categ.* **6**: 83–93.

Rezk, C. W. (1996). Spaces of algebra structures and cohomology of operands. Thesis, MIT.

Roos, J. E. (1965). Caractérisation des catégories qui sont quotients des modules par des sous-catégories bilocalisantes. *C. R. Acad. Sci. Paris* **261**: 4954–4957.

Rosický, J. (2007). On homotopy varieties. *Adv. Math.* **214**: 525–550.

Rosický, J., and E. M. Vitale (2001). Exact completion and representations in abelian categories. *Homology Homotopy Appl.* **3**: 453–466.

Schubert, H. (1972). *Categories*. Springer.

Street, R. (1974). Elementary cosmoi I. In *Category Seminar Sydney*. Lect. Notes in Math. **420**. Springer, 75–103.

Street, R., and R. F. C. Walters (1978). Yoneda structures on 2-categories. *J. Algebra* **50**: 350–379.

Tholen, W. (2003). Variations on the Nullstellensatz, paper presented at the European Conference on Category Theory, Haute Bodeux, Belgium.

Ulmer, F. (1968). Properties of dense and relative adjoints. *J. Algebra* **8**: 77–95.

Vitale, E. M. (1996). Localizations of algebraic categories. *J. Pure Appl. Algebra* **108**: 315–320.

Vitale, E. M. (1998). Localizations of algebraic categories II. *J. Pure Appl. Algebra* **133**: 317–326.

Voronov, A. A. (2005). Notes on universal algebra. In *Proc. Symp. Pure Math 73*. AMS, 81–103.

Watts, C. E. (1960). Intrinsic characterizations of some additive functors. *Proc. Amer. Math. Soc.* **11**: 5–8.

Wechler, W. (1992). *Universal Algebra for Computer Science*. Springer.

Wraith, G. C. (1970). *Algebraic Theories*. Aarhus Lecture Note Series **22**. Aarhus Universitat.

List of symbols

$\mathcal{A}^\mathcal{C}$, 4
$\mathcal{A}_{fp}$, 62
$\mathcal{A}_{pp}$, 59
$(\mathcal{A}, U)$, 9
ALG, 84
ALG^1_u, 236
ALG^1, 114, 236
ALG^S, 142
ALG_{colim}, 161
Ab, 12
Add, 228
Alg, 85
$Alg\ M$, 80
$Alg\ T$, 105
$Alg\ \mathcal{T}$, 11
Alg^1, 114, 239
$Bool$, 59
E_{Ic}, 75
E_{Ind}, 43
E_{Rec}, 65, 191
E_{Sind}, 43
E_{Th}, 15
$E_\mathbb{D}$, 40
$El\ A$, 8
Ex, 174
$FMon$, 223
F_H, 120
F_T, 109
F_Σ, 120, 127
$F_\mathbb{M}$, 212
$Graph$, 14
H-Alg, 28, 117
H_Σ, 28
$Ic\ \mathcal{C}$, 75
$Ind\ \mathcal{C}$, 43
$J^f_\mathbb{M}$, 222

$J_\mathbb{M}$, 219
$K_\mathbb{M}$, 219
LEX, 87
LFP, 87
$Lc\alpha$, 174
Lex, 87
$Lex\ \mathcal{T}$, 44
M-set, 132
Mon, 34
Pos, 29
$PsALG^1$, 237
$PsTh^1$, 236
R-Mod, 13
$RGraph$, 107
$Rec\ \mathcal{C}$, 65, 190
S-sorted signature, 14
S^*, 139
Set, 3
Set^S, 12
$Set^\mathbb{M}$, 215
$Set^f_\mathbb{M}$, 222
$Sind\ \mathcal{C}$, 43
T_Σ, 128
Th, 84
Th^1, 114, 236
Th^S, 142
Th_c, 84
Th_{bim}, 161
U_H, 117
U_Σ, 127
$U_\mathbb{M}$, 210
$Y_\mathcal{C}$, 4
Δ, 7
Γ, 174, 183
Φ_A, 8
Σ, 14

$\Sigma\text{-}Alg$, 14
$\mathbb{M}$, 208
$\mathcal{C}(\mathcal{T}, T)$, 129
$\mathcal{C}^{op}$, 4
$\mathcal{G}$, 54
$\mathcal{K} \downarrow K$, 7
$\mathcal{K}^{\mathbb{M}}$, 210
$\mathcal{K}_{\mathbb{M}}$, 219
$\mathcal{N}$, 12
$\mathcal{T}$, 11
$\mathcal{T}/\sim$, 92
$\mathcal{T}^{[k]}$, 154
$\mathcal{T}_\Sigma$, 128

$\mathcal{T}_{ab}$, 12
$\mathcal{T}_C$, 15
$\mathcal{T}_{Mon}$, 109
uTu, 155
$\mathcal{P}'_{ex}$, 185
$\mathcal{P}_{ex}$, 174, 182
$\dashv$, 6
π_i^n, 103
$(F \downarrow G)$, 7
$(\mathcal{A}, U)$, 9
$(\mathcal{T}, T)$, 104
2-category, 82
2-functor, 83

Index

Abelian category, 227
absolute coequalizer, 214
absolutely presentable object, 48
additive category, 227
additive functor, 227
adjoint functor theorem, 6
adjoint functors, 6
algebra for a signature, 14, 142
algebra for a theory, 11
algebra for an endofunctor, 28, 117
algebraic category, 11
algebraic functor, 81
algebraic theory, 11
amnestic concrete category, 133
arity, 14

bicategory, 161
biequivalence of 2-categories, 84
bimodule, 160
biproduct, 227
Birkhoff's variety theorem, 96

canonical algebraic theory, 78
category of algebras for a signature, 14
category of algebras for a theory, 11
category of algebras for an endofunctor, 28
chain, 22
clone, 147
co-well powered category, 33
cocomplete category, 4
comparison functor, 213
complete category, 4
concrete category, 9
concrete equivalence of categories, 106

concrete functor, 9
concrete isomorphism of categories, 134
concretely equivalent categories, 106
concretely isomorphic categories, 134
congruence on a theory, 90
congruence on an algebra, 112
connected category, 24
conservative functor, 4
cospan, 25
counit of an adjunction, 6

dense, 57
diagonal functor, 7
diagram, 4
directed colimit, 22
directed graph, 14
directed union, 23, 51

effective equivalence relation, 35
Eilenberg–Moore algebra, 210
Eilenberg–Moore category, 210
enough regular projective objects, 50
equation, 89, 130, 144
equational category, 131
equational theory, 131
equivalence of categories, 3
equivalence relation, 34
equivalent objects in a 2-category, 84
essentially surjective functor, 3
essentially unique extension, 40
exact category, 35
exact functor, 167
extremal epimorphism, 9

faithful functor, 3
filtered category, 5, 21

filtered colimit, 21
filtered colimit in *Set*, 5
filtered diagram, 5, 21
final functor, 24
finitary functor, 28
finitary localization, 199
finitary monad, 208
finite S-sorted set, 16
finite diagram, 4
finitely generated Σ-algebra, 136
finitely generated category, 4
finitely generated free algebra, 109, 142
finitely generated object, 51
finitely presentable Σ-algebra, 136
finitely presentable object, 47
free Σ-algebra, 127
free H-algebra, 120
free algebra, 109, 142
free completion under $\mathbb{D}$-colimits, 40
free completion under colimits, 41
free completion under filtered colimits, 44
free completion under finite products, 15
free completion under reflexive coequalizers, 65
free completion under sifted colimits, 43
free Eilenberg–Moore algebra, 211
free exact completion, 174
free monad, 210
free theory on a signature, 129
full functor, 3

generator, 54

homomorphism of Σ-algebras, 14, 142
homomorphism of H-algebras, 28
homomorphism of algebras, 11

idempotent complete category, 74
idempotent completion, 75
idempotent modification, 155
idempotent morphism, 22
initial H-algebra, 117
isomorphism of categories, 4

kernel congruence, 91
kernel pair, 8
Kleisli category, 219

left-covering functor, 167
localization, 199
locally finitely presentable category, 61

matrix theory, 154
monad, 208
monad induced by an adjunction, 209
monad morphism, 217
monadic category, 214
monoid, 34
Morita equivalent algebraic theories, 154
Morita equivalent monoids, 158
Morita equivalent ringss, 153
morphism of S-sorted algebraic theories, 140
morphism of algebraic theories, 80
morphism of one-sorted algebraic theories, 104
morphism of signatures, 128

one-sorted algebraic category, 106
one-sorted algebraic theory, 104
one-sorted signature, 14

perfectly presentable object, 47
polynomial functor, 28
preadditive category, 227
pseudoconcrete equivalence, 232
pseudoconcrete functor, 232
pseudoconcretely equivalent categories, 232
pseudomonadic category, 232
pseudomorphism of one-sorted theories, 234
pseudo one-sorted algebraic category, 233
pseudoequivalence, 169
pseudoinvertible morphism, 154

quasi-inverse functor, 3
quasivariety, 99
quotient functor, 124
quotient object, 8

reflective subcategory, 7
reflexive coequalizer, 30
reflexive pair of morphism, 30
reflexive relation, 34
regular epimorphism, 8
regular epireflective subcategory, 95
regular factorization, 33
regular factorization of a parallel pair, 169
regular projective algebra, 46
regular projective cover, 166
regular projective object, 46
relation, 34
representable algebra, 15

representable functor, 7
retract, 8

S-sorted algebraic category, 141
S-sorted algebraic theory, 140
S-sorted equational category, 145
S-sorted set, 12
sifted category, 21
sifted colimit, 21
sifted diagram, 21
signature, 14
slice category, 7
solution set, 6
split epimorphism, 8
splitting of an idempotent, 74
strong epimorphism, 8
strong generator, 54
subalgebra generated by a set, 106
subobject, 9
symmetric relation, 34

term for a signature, 127
transitive relation, 34
transportable concrete category, 133
tree, 119

unary theory, 159
uniquely transportable concrete category, 133
unit of an adjunction, 6

variety, 91

weak limit, 167
well-powered category, 197
word, 16

Yoneda embedding, 4
Yoneda lemma, 4

zero object, 227

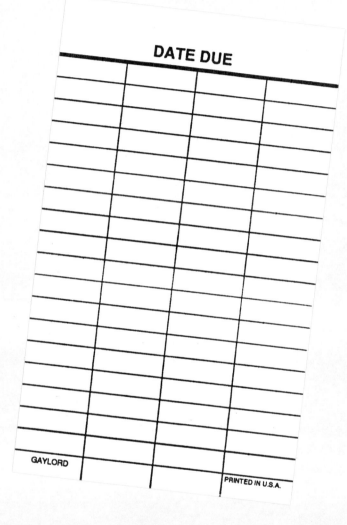

```
QA 169 .A31993 2011
Adámek, Jiří,
Algebraic theories
```